READINGS IN INNOVATIVE IDEAS IN TEACHING COLLEGIATE MATHEMATICS

Edited by
Mohammad H. Ahmadi

Introduction by
Thomas A. Romberg

D1417346

University Press of America,® Inc.
Lanham · New York · Oxford

Copyright © 2002 by
University Press of America,® Inc.
4720 Boston Way
Lanham, Maryland 20706
UPA Acquisitions Department (301) 459-3366

12 Hid's Copse Rd.
Cumnor Hill, Oxford OX2 9JJ

ISBN 0-7618-2325-5 (paperback : alk. ppr.)

To Homa, Hoda and Reza

CONTENTS

SECTION 1
COOPERATIVE LEARNING

SECTION 2
EDUCATIONAL TECHNOLOGY

SECTION 3
ATTITUDE

PREFACE

This book deals with the practical aspects of innovative ideas in teaching mathematics, but neither the theoretical nor historical viewpoints of any particular pedagogical approaches. This volume is formed of a collection of papers contributed by professional mathematicians in United States and Great Britain. The topics in the book represent the pros and cons of implementing creative instructional styles in classrooms. The main objective of this book is to share the implementation and outcomes of these experiments with the teachers of mathematics at all educational levels and help them to learn about different teaching strategies and how they might incorporate in their classrooms.

Mohammad H. Ahmadi
University of Wisconsin-Whitewater

ACKNOWLEDEMENTS

I would like to thank professor Thomas A. Romberg of University of Wisconsin-Madison for his assistance in selecting the papers. His introduction to this volume made it a solid and valuable book to read. I wish also to thank my son — Reza Ahmadi — for his assistance in typing and reconstructing some of the figures. Finally, my sincere thanks go to contributors whose collaboration resulted in the book.

INTRODUCTION

Thomas A. Romberg

Collegiate teachers of mathematics often face a complex and difficult problem. Many students in their classes are required to take mathematics course, but lack the background, inclination, or enthusiasm needed to succeed. Although these students have completed high school courses in algebra and geometry, they failed to learn well the concepts and skills in those courses. They only learned procedures that were needed to pass tests, or in the interim years since their last mathematics course they forgot what had once been learned because it had not been learned with understanding. There are many causes underlying this problem. Nevertheless, collegiate teachers must deal with such students in their classes. Instruction in classes with such students can be frustrating, but many dedicated teachers have made interesting attempts to change their instructional practices that have resulted in improved student learning and attitudes.

In this volume Professor Mohammad H. Ahmadi, University of Wisconsin — Whitewater, has collected a set of papers written by teachers about instructional practices that are promising and certainly worth considering by others. The papers are grouped into three sections. The three papers in the first section are on cooperative group practices. Too often mathematics teachers have not discouraged students from working and studying in cooperative groups, sharing strategies, clarifying one's notions with others, etc. One of the papers report on Ahmadi's own work in finite mathematics and business calculus classes in which cooperative learning was encouraged with traditionally taught courses. The second paper in this section spirals cooperative group discussions of pedagogy through the core mathematics courses for secondary mathematics teachers. The third paper reports the impact of cooperative learning on students problem-solving behaviors and achievements in a college algebra and statistics course. The cited evidence in these papers make it clear that working in cooperative groups motivates many students to achieve and change their attitudes about mathematics.

The second section contains three papers on the use of technology in the teaching of mathematics course. In the first paper instructors used video tapes as ancillary materials for two units in a college algebra course. The instructors and students benefited from the opportunity to replay explanations and examples. However, it is interesting to note that the instructors were "especially enthusiastic about using collaborative learning in class", a feature used to have students discuss the videos. The author of the second paper reports on the use of spreadsheets to convey the dynamic aspects of limiting processes studied in calculus. The final paper in this section is a description of a computer based diagnostic testing system used in two universities in Great Britain.

The third section contains five papers where the focus is changing student perceptions and attitudes toward mathematics. Two papers describe mentoring support systems for students in calculus classes. Then two papers outline a variety of tactics used to provide students with interesting and challenging problems. The authors of one paper argue too often math problems are viewed by students as "alien" (i.e. having little relevance to anything in their world). The final paper describes a component system for college courses that allows students to progress via segments of typical courses.

In summary, this collection of papers directly address a major problem that collegiate mathematics teachers face — too many students in their classes lack the background, inclination, or enthusiasm needed to succeed. The authors of these papers make it clear that such students can succeed but they need support. Taking a mathematics course that covers material taught earlier without understanding at a faster pace dooms many to failure. However, if one provides support via cooperative group learning, new technologies, more relevant tasks, mentoring, etc., many students change their views of mathematics and achieve.

SECTION 1
COOPERATIVE LEARNING

1

THE IMPACT OF COOPERATIVE LEARNING IN TEACHING MATHEMATICS[†]

M. H. Ahmadi

Abstract

This report presents an analysis of the results of an experimental study conducted by the author at the University of Wisconsin — Whitewater in two mathematics courses: one section of Finite Mathematics, and two sections of Business Calculus. Both courses are designed for students majoring in business and social sciences. The experiment involved a nontraditional style of teaching, a combination of lecture and group discovery methods. The approach involved considerable student interaction, both in and out of the classroom; it used a team format and had less formal instruction. The dependent variables in the study were student performance, interest, motivation, conceptual understanding, and attitudes toward mathematics. To determine the effectiveness of the method of instruction on the dependent variables, comparisons were made with data from other sections of the same courses I taught using more traditional methods. The results of the study demonstrated that students in the study performed better than those in traditional sections; their attitude towards mathematics improved; they actually participated in outside classroom work and became more interested in and motivated to do mathematics; and the majority of the students were positive about the instructional approach and thought this method was a better way for them to learn mathematics.

Introduction

I have taught mathematics for over two decades. Throughout that time I have been trying to motivate students and change their attitudes towards mathematics. I tried different instructional strategies and considered several factors learned through personal experiences and readings. Although I was somewhat successful in achieving my goals, I was not really satisfied with the way I taught mathematics because, as with most mathematics instructors, I relied on well-presented lectures. Unfortunately, I found that students who are supposed to be the center of the learning process were quiet and developed a passive attitude in my math classes. To be fair, this method is not bad if the class size is small because considerable student/teacher interaction is possible. But, due to economic reasons, most university mathematics courses are quite large. This, in turn, leads to little student/teacher interaction.

The experimental approach used in these studies was designed to complement lectures with structured small-group work. The design of the approach I used was an outgrowth of prior experiences. For example, in the spring of 1989 I decided to use small-group work in teaching a section of "Geometry for Elementary Teachers." I found that the instruction went very well, and the students were very positive about this approach. However, as a consequence of this experience, the need for detailed plans became clear. For example, in the geometry class there were only 24 students. This made it possible for me to organize and monitor student work, but in another course with 40 students I did not apply this approach because I could not structure the group learning approach for that many students. Nevertheless, from these experiences I was convinced that an instructional approach utilizing small-group work was viable and more effective than the traditional lecture approach.

Design of the Experiment

During spring 1991, I decided to design a systematic experiment using a small-group learning approach in teaching one section of "Finite

Mathematics" (Math 143) and two sections of "Business Calculus" (Math 243) in the fall of 1991. I spent a good portion of the summer designing a way of implementing a group learning method in these classes. Some of the questions considered in the design were: What should the evaluation system be (individually and in groups)? On what topics should lectures be given? How much information should be revealed? What part of a lecture should be omitted for students' exploration? What topics/concepts should the students explore? How can a teacher control the students' outside group activities? How should a teacher form small groups? The experimental treatment grew out of my attempts to respond to these and other questions. What follows is a summary of the structure of my approach, with respect to nine issues:

1. Forming small groups. On the first day of class, I distributed 3 x 5 cards to the students and asked them to provide names, telephone numbers, previous mathematics courses taken in high school and college, and the grade received in each course. After collecting this information, copies of the course syllabus were distributed, and students were asked to read them very carefully. I wanted them to be absolutely aware of what they would be expected to do if they chose to stay in my class. Based on the information provided on the cards, I then divided the class into seven nonhomogeneous groups of five members each. The information I considered was past performance in mathematics, gender, and race. For past performance, I assumed that good letter grades (A or B) in prior courses indicated good mathematical ability. However, I was well aware that this assumption is sometimes false (e.g., sometimes a "C" student from one high school performs better in college mathematics than an "A" student graduating from another high school).

2. Appointing a group leader. I chose one student in each group to be the leader of his/her group. Also, because there was a possibility that a leader might dominate the discussion in a group the following points were made to the entire class:

- The leader of a group is not necessarily the best math student in the group.
- There are no incentives (such as extra points) for leaders.
- Leaders and members of their group will be treated equally.

3. Creating a Seating pattern. Members of each group were asked to sit close to each other so that they could quickly rearrange their seats for group activities. On the first day of class, leaders were asked to get together with their members and, after introductions, take names and phone numbers of their group members and arrange their first outside meeting. I wanted to save class time and have them to set up their weekly meeting schedule for outside classroom activities. The first meeting was to take place before the next class session so that I could have their schedules to make sure they could meet for at least five hours per week.

4. Organizing Outside classroom activities. Leaders and group members were given the following instructions:

- Each member of the group was to study his/her notes and the related section(s) in the textbook before attending group session.
- They were to write down any unanswered questions they had.
- At each meeting, they would first spend a few minutes for a question-answer discussion on current topics.
- For each assigned problem, after exchanging ideas and discussing different possible strategies, each member would start to solve the problem. Then members would compare and analyze results and make corrections or changes.
- No member or leader of a group should rush to provide a solution to a problem before all members of the group had a fair chance to work and to discuss their approaches to solving the problem.
- No leader or member of a group should dominate the discussion.
- Each group would hand in one set of solutions to homework assignments from sections that were covered the previous week. A statement of each problem and its solution was to be written clearly and legibly on one side of the paper. On the top of the first page they were to write their group number and the names of all group members. The final draft of the solution set was to be reviewed by all members.
- Each member of the group would take turns writing the solution sets. (NOTE: Some leaders asked me whether the members of their groups could share the writing of the solution sets, rather than one member being responsible for writing up the whole solution set. I gave them permission to do so.)
- If the group couldn't solve a problem, one of the group members could come to me or go to the Math Tutoring Center and ask for help.

• They were to show all their work in their solutions.
• If there were any problems or issues concerning a member of the group, they could reach me at my office or at home. They were not to wait until the next session for the issue to be resolved.

5. Monitoring outside classroom activities. I had copies of the weekly meeting schedules (day, time, and place of the meetings). I then visited the groups at random, without notice. All groups had their meetings at the University Center and I made sure to visit each group at least once a week. The main reason for paying such visits was to make sure that everything was going as planned and that students were taking their assigned tasks very seriously. Furthermore, from such visits, combined with comments from leaders, talking to students on an individual basis outside the classroom, meetings with each group after each test regarding each member's performance on the test, and invitations to those few students who performed poorly to my office for individual consultation, I was able to detect and correct problems within one or two of the groups.

6. Designing an evaluation scheme. Each student's course grade was based on the following two factors: (a) Group Evaluation: 30% (homework assignments) = 75 points and (b) Individual Tests: 70% (three tests and a final) = 175 points. The total available points were 250. I used a 90 - 80 - 70 - 65 percent scale for letter grades "A," "B," "C," and "D." The distribution of grades for each class was the indicator of student performance I used as the initial dependent variable in the three classes.

7. Selecting an instructional method. A combination of lecture and group-discovery approach (with less emphasis on lecturing) was used. In the lectures I provided enough (but not complete) information in class on various topics and discussed each in enough detail so that the groups could work effectively on specific problems.

Group activities during the class began by raising questions (if needed) with the entire class. My goal was to let students reformulate the definitions, express the mathematical ideas in their terms, and explore strategies to solve particular problems. Sometimes I used a portion of or occasionally the entire class period for group activities. During this time I moved from one group to another providing assistance, or encouraging use of different approaches to solving a math problem.

8. Preparation of instructional materials. I reviewed and revised each course syllabus in order to be sure that main topics were covered during the semester. In this regard I made several changes, such as the elimination of topics overlapping within the textbook, or with prior courses taken by the students. For example, in business calculus the topics on break-even analysis (already taught in Finite Mathematics) and graphing quadratic functions using algebra were eliminated. Finally, I reviewed the text of each course carefully to decide how the materials should be presented, and which questions should be used for group activities. In short, I decided for some topics what should be lectured on and what the students should rediscover. For example, after introducing the Simplex algorithm using a two-variable standard linear programming problem, I asked each group to do the same problem using a graphical approach. Their next task was to compare their graphs with completed simplex tableaus. All groups realized that each simplex pivot will take us from one corner point of the constraint set to an adjacent one, where the value of the objective function is greater, and the process terminates when the optimum is achieved. Therefore, they rediscovered the geometric meaning of simplex method. The same was done for nonstandard case, and students understood that why this case involves two phases.

9. Conducting student surveys. I conducted two surveys, one at the beginning and the other at the end of the semester. The first survey contained two scales: "Students' Conception of Mathematics," and "Students' Attitude Toward Mathematics." The second survey had an additional scale "Students' Evaluation of Group Learning Approach." The "Students' Conception of Mathematics" scale was adapted from Romberg, et al. [2] and comprised six statements, which students were to judge on a 5-point Likert scale with respect to the degree to which they felt the statement reflected their own conception of mathematics. A sample statement was "Mathematics is a language, with its own precise meaning and grammar, used to represent and communicate ideas."

The "Students' Attitude Toward Mathematics" scale included three statements: "I like mathematics"; "I dislike mathematics," and "I hate mathematics."

The "Students' Evaluation of Group Learning" scale included eight statements about group learning. The statements were adapted from those written by D. W. Johnson and R. T. Johnson (referenced in Davidson, [1]). Again, each statement was to be judged on using a 5-point Likert scale. A sample statement was, "Group learning motivated me to learn and explore mathematics better." Copies of these surveys are included in Appendix A.

Conduct of The Study

Sixty-nine students enrolled in two sections of Math 243 (Business Calculus) and 35 students in one section of Math 143 (Finite Mathematics) at the University of Wisconsin-Whitewater participated in the study. There are many sections of these courses offered each semester. When students enrolled in these sections, they did not know they were experimental sections. The students became aware of the nature of the experiment only on the first day of class and were given the opportunity to switch sections. The sections met on Tuesdays and Thursdays for 75 minutes. In the sections, seven groups of four or five students each were formed. During the semester my activities included, but were not limited to, a brief lecture, usually at the beginning of each class session; monitoring and observing group problem-solving both in and out of class; interviewing students and leaders outside class; reviewing weekly homework assignments which were graded by assistants under my direction; administering surveys; and grading all exams.

Preliminary Evaluation

During the semester two preliminary evaluations of the experiment were conducted.

(a) Because I wanted to know whether this approach was working, I asked the students during the fourth week of classes to express their opinions about the group-learning method on a short questionnaire. Overall, the students were very positive. Some typical responses are given below: I think the group meetings are working really well. We help each other with things we don't understand. Sometimes it is easier to ask a peer for help than to approach your professor. I'm not sure if I understood this test material because of the group or because it was review, but I like the group idea.

I think it is an excellent approach to studying calculus. I am taking this for the second time, and I have learned more in the first few weeks than I did all year last year.

I love the study group method. It usually turns out that I have the answer to someone's question, and they have an answer to my question. I would like to keep this method and my current study group. It is a big help!

Last semester I took Math 243 for half a semester and then dropped it. I was totally lost, without a clue. This semester I feel very confident about my work in this class. I think the reason for that is because now I have people I can ask for help and get their interpretations of the lessons. I was dreading this class in the beginning, but I am actually enjoying it because I know what I'm doing.

The effects of study groups have helped me personally. I am able to get my questions answered on a one-to-one basis. I enjoy the study group, and math has been more enjoyable.

I really like working in a group outside of class. That way if you don't understand something you can be taught right there. Plus people aren't afraid to say that they don't understand something and they might be afraid to speak up during class, so it's a great idea.

(b) The second preliminary evaluation was done by my colleague Dr. Ronald Detmers who visited my Math 243 (Business Calculus) class before the end of the semester. He observed group activities in the classroom while each group was working on an assigned problem. Dr. Detmers made the following comments:

During this portion of the class period I went around to the various groups and visited. Particularly I asked them how they liked this method of studying Mathematics. I found most of the responses were positive. For example:

'I am repeating this course and I find it much easier with a support group to ask for help and also to give help.'

'It keeps us working on assigned problems that as individuals we would have given up on because we don't want to let the group down.'

All in all I feel it was a positive situation, the students were enthusiastic about the materials, They were definitely positive about the experimental approach and seemed to be learning the material.

Results

To determine the effectiveness of the experiment, data were collected on student performance (grades and percent of drops), conception of mathematics, and attitudes toward mathematics

Distribution of Grades

Comparisons were made on grade distributions and dropouts with sections of the same courses taught using traditional method, with the same textbook and exams. Table 1 shows the percentage rate of students' grades. The first row indicates the overall averages of Math 143 that I taught traditionally in the past. The second row shows the results of the study. The same was done for Math 243. A quick inspection of this table makes it obvious that the distribution of grades was considerably higher in all three of the experimental sections. Of particular note is the sharp reduction of Ds and Fs in the experimental sections.

Dropouts

The percent of students that dropped each experimental class was also compared to the drop rate for traditionally-taught sections of each course. For Math 143, the average drop rate was 15.7%, but for experimental section only 2.9%. Similarly for Math 243, the average drop rate was 15.9%, but 2.9% and 0% for the two experimental sections.

Along with the reduction in low grades, these data on the significant reduction in percent of students who dropped the courses imply that those students who needed help were getting it as a consequence of the group support. It should be noted that no students in the experimental sections left on first day of classes to transfer to another section.

Conception of Mathematics

Students in the experimental sections were asked to complete the survey on conceptions of mathematics: Survey (I) at the start and Survey (II) at the end of the experiment. The six statements in the survey portray a variety of viewpoints on the nature of mathematics. Students rated each statement on a 5-point Likert scale. The percent of all students ($\underline{N}$ = 102) responding to each statement for both surveys is shown in Table 2. The numbers appearing in the first row indicate the percentage of student responses from Survey (I) and the ones in the second row are from Survey(II). The data on the students' shift in ratings on the first two statements indicate a positive shift toward mathematics as a language and a process to solve real-world problems as a consequence of the course. However, there was no discernible shift by the students with respect to the other four more traditional conceptions. This implies that the students simply expanded their conceptions of mathematics rather than changing them.

Attitude Toward Mathematics

On each survey students were asked whether they liked, disliked, or hated mathematics. The responses for each of the sections are shown in Table 3. The positive shift (from 40 to 60%) in attitudes toward liking mathematics was apparent in both sections of the Business Calculus course. Also, because males and females often have differing attitudes toward mathematics, I examined if the positive shift observed in these classes was similar for both groups (see Table 4). This data indicates that the shift in attitude of males toward mathematics was primarily from dislike to like, while for females it was away from hating mathematics.

Group Learning Approach

At the end of the semester I asked the students to evaluate the group learning approach used in the course on eight anticipated student outcomes. Table 5 gives the percent of students that agreed that the approach had helped them on that outcome or were neutral.

Clearly a high percentage of students agreed that the Group Learning Approach helped them to work more productively on mathematics problems. It should be noted that the percentage rate of those students who disagree that the method helped them was left out, but the rate of disagreement was at most in the lower teens.

Follow-up Interviews

During the next semester 18 students who had been in the experimental section of Math 143 were then taking Math 243 with me. The course was taught using a cooperative learning approach. I interviewed these students about their experience in the experimental course. These are examples of the comments they made (all said that group learning worked better for them):

• Easy to ask questions from peers and discuss the problems.

• In the past, if I was not good in one section, I just skipped that section. But this did not happen this semester.

• I was not frustrated.

• If I got stuck, there was someone to help.

• I usually didn't do homework until the test comes.

I didn't like it at first, but then I did. [I asked why.] Because I thought that we couldn't meet all together outside the classroom on a regular basis.

Conclusion

The results of this study make it clear that there are several advantages of the group learning approach in teaching mathematics courses as compared to more a traditional method of instruction. First, this approach is very dynamic. Students are more actively involved in working out specific problems or exploring some of the mathematical concepts and ideas. Second, the students really do mathematics. This is the part of the learning process that the lecture method does not automatically achieve. Most students think that knowing how to do a specific mathematics problem suffices and hence get frustrated when they cannot work out the problems and get the correct answers on tests, even though they were able to do a few specific example problems. Third, this approach provides students the opportunity to communicate mathematics. In fact, each member of the group has to read his/her notes and the related sections before attending inside or outside the class group meetings, write up weekly solution sets, and then discuss mathematical ideas and problems during group meetings in or outside the classroom. Fourth, students learn to cooperate with other members of their groups. Fifth, students learn and practice group decision-making, an important element in today's sophisticated technological society. Sixth, the teacher-student relationship is closer than in traditional approach. Seventh, the classroom atmosphere is friendly and less formal so that students feel free to ask questions, get involved in group discussions, and respond to my questions without being afraid of being wrong. Finally, close contact with students during their group activities inside or outside the classroom enabled me to see how each student was thinking and approaching a particular problem.

Overall, based on this and several other experimental studies I have conducted, I consider the experiment to be a resounding success, and I would strongly advocate that other mathematics instructors try the approach. I have been implementing cooperative learning in my classes for the past few years with some modifications. For example, in second semester Calculus I used cooperative learning method with group research projects as an added feature; weekly outside classroom activities were recommended but were optional. However, anyone considering this method must take in account certain points. First, it is more time consuming (because of preparation, monitoring, etc.) than the traditional lecture method. Second, there may be scheduling conflicts for regular meetings outside the classroom, but this issue can usually be resolved by reassigning student(s) to a different groups. I had only one student who was unable to make any outside classroom meeting, and hence it was not feasible to assign him to a group. I gave the student two options: to help him transfer to another section or to work all the weekly homework assignments by himself (he chose the latter). Third, inevitably some students are resistant to studying mathematics in a new way. This was particularly true for some very good students who felt the group meetings involved spending more time than necessary, but they began to like it as the semester progressed. But subsequent studies showed me that this situation can be avoided. On the first day of class I simply say "some good students may think that working in groups and sharing their knowledge is a waste of time, but I am convinced that explaining mathematics to your peers will help you to understand mathematical ideas even better." Fourth, this approach is slow at the beginning by its very nature, because for most students this is probably their first experience in learning and doing mathematics in a group setting and they need time to adjust to a new instructional style. Usually the coverage of topics will reach its normal pace in about two weeks.

Acknowledgments

The Department of Mathematics and Computer Science, University of Wisconsin — Whitewater, supported this experimental study. However, the opinions, findings, and conclusions are those of the author only.

I am indebted to Professor Thomas Romberg for giving me his advice during the course of this study and his helpful comments on the draft of this paper. Finally, my sincere thanks go to my colleague, Professor Robert Knapp, for reviewing this paper.

Notes

1. Although all the leaders had grades of A or B in high school, three leaders received a grade of C and some received Bs in the class even though some members of their groups received As. Out of the 21 leaders, 11 got As, 7 got Bs, and 3 got Cs.
2. The "Students' Conception of Mathematics" scale was adapted from Romberg [2].

3. The statements in the "Students' Evaluation of Group Learning" scale were adapted from those used by Johnson and Johnson (referenced in Davidson [1]).

† Reprinted with permission from PRIMUS-Problems, Resources, and Issues in Mathematics Undergraduate Studies. Vol. X, No. 3, pp. 225-240 (2000). Published at the United States Military Academy, West Point NY.

References

Davidson, N, 1990. Cooperative learning in mathematics. Reading, MA: Addison-Wesley.
Romberg, T. A., Middleton, J. A., Webb, N. L., & Pittelman, S. D. 1990. Teacher's conceptions of mathematics and mathematics education. Wisconsin Center for Education Research, School of Education, University of Wisconsin-Madison.

Table 1: Students' Performance (Grades)

	A	B	C	D	F
MATH 143					
Traditional	30.0	32.0	19.0	3.0	16.0
Experimental	52.9	32.4	14.7	0.0	0.0
MATH 243					
Traditional	17.0	25.0	27.0	12.0	19.0
Experimental					
Section 02	24.2	42.4	21.2	6.1	6.1
Section 05	27.3	42.4	24.2	0.0	6.1

Table 2: Students' Conception of Mathematics

Statement	Survey	A/SA	N	D/SD
Mathematics is a process in which abstract ideas are applied to solve real-world problems.	I	67.4	22.5	10.1
	II	84.8	13.0	2.2
Mathematics is a language, with its own precise meaning and grammar, used to represent and communicate ideas.	I	80.9	16.9	2.2
	II	93.4	5.4	1.1
Mathematics is a collection of concepts and skills used to obtain answers to problems.	I	97.8	2.2	0.0
	II	98.9	1.1	0.0
Mathematics is thinking in a logical, scientific, inquisitive manner and is used to develop understanding	I	87.6	10.1	2.3
	II	89.3	9.7	1.0
Mathematics is facts, skills, rules and concepts learned in some sequence and applied in work and future study.	I	74.2	21.3	4.5
	II	82.6	14.3	3.3
Mathematics is an interconnected logical system, is dynamic, and changes as new problem-solving situations arise. It is formed by thinking about actions and experiences.	I	75.3	20.3	12.4
	II	66.3	28.3	5.4

NOTE: "A/SA" stands for "Agree or Strongly Agree," "N" for "Neutral," and "D/SD" for "Disagree or Strongly Disagree"

Table 3: Students' Attitude Toward Mathematics

Courses	Section	Like	Dislike	Hate
Math 143	I	82.8	06.9	10.3
(N = 34)	II	88.0	09.0	03.0
Math 243 #2	I	45.1	42.0	12.9
(N = 33)	II	60.6	36.4	03.0
Math 243 #5	I	37.0	48.0	15.0
(N = 35)	II	59.3	33.3	07.4

Innovative Ideas in Teaching Collegiate Mathematics

Statements (Tables 3 and 4): 1. I like mathematics; 2. I dislike mathematics; 3. I hate mathematics.

NOTE: It would have been better to give students more choices by adding a statement such as "None of the above."

Table 4: Male and Female Attitudes Toward Mathematics

Course	Survey	Like	Dislike	Hate
Males	I	90.0	00.0	10.0
	II	92.3	00.0	07.7
Females	I	79.0	10.5	10.5
	II	85.0	15.0	00.0
Math 243 #2				
Males	I	37.5	50.0	14.3
	II	57.2	35.7	07.1
Females	I	52.9	35.3	11.8
	II	63.2	36.8	00.0
		Math 143		
Math 243 #5				
Males	I	40.0	53.0	07.0
	II	70.6	23.5	05.9
Females	I	33.0	42.0	25.0
	II	40.0	50.0	10.0

Table 5: Students' Evaluation of Group Learning Approach

Statement	Math 143 (N = 34)		Math 243 #2 (N = 33)		Math 243 #5 (N = 35)	
	A/SA	N	A/SA	N	A/SA	N
1	60.6	27.3	61.8	20.6	88.9	7.4
2	72.8	9.1	67.7	14.7	92.7	3.7
3	63.6	18.2	67.7	11.8	85.1	11.1

	A/SA	N	A/SA	N	A/SA	N
4	39.4	42.4	50.0	26.5	55.5	40.7
5	48.5	45.5	55.9	20.6	85.2	11.1
6	60.6	21.2	52.9	26.5	74.0	14.8
7	72.7	21.2	61.8	26.5	85.2	11.1
8	48.5	42.4	61.8	26.5	70.3	22.2

Statements

Group learning:

1. Motivated me to learn and explore Math better.
2. Helped me to be a better Math problem solver.
3. Accelerated my math performance and achievement.
4. Changed my attitude toward math for the better.
5. Built confidence in my ability to reason mathematically.
6. Gave me the ability to accept frustration when I couldn't figure out problems.
7. Gave me the willingness to persevere when solutions are not immediate.
8. Lead me to attribute slow progress in finding answers to not using the right strategy rather than not being competent.

NOTE: "A/SA" stands for "Agree or Strongly Agree," and "N" for "Neutral." Table 5 does not show the percentage rate of "Disagree or Strongly Disagree," but it can be found easily from 100 — (A/SA + N).

2

A CASE STUDY OF SPIRALING CONTENT AND PEDAGOGY THROUGH CORE COURSES FOR PRE-SERVICE SECONDARY MATHEMATICS TEACHERS

Tabitha T. Y. Mingus

Introduction

Mathematics teachers employed by schools and universities are touched by reform efforts, whether or not they personally agree with the philosophies driving reform efforts or are involved in implementing reform-based curricula. The influence of reforms can range from a limited awareness of the changes underway, to complete immersion into reform-minded approaches to teaching mathematics, to a vocal disagreement with the assumptions behind the reform trends.

Despite the traditional "academic freedom" enjoyed by individual teachers, the inertia behind the current reform movements can be felt in classrooms whether or not the teacher agrees with their basic precepts. The textbooks available, mandated district curricula, fellow teachers' opinions and the public's expectations of mathematics teachers are all influenced by the current pendulum swing supporting the use of a reformed mathematics curriculum. These influences in addition to dwindling resources and bulging class sizes can alter the manner in which reform is implemented from school to school and classroom to classroom (Ferrini-Mundy & Johnson, 1997).

Mathematics educators at the university level who are responsible for training pre-service teachers need to be cognizant of the pressures and responsibilities their students may face. Pre-service mathematics teachers need to be well-informed about the historical development and use of reform methods (Bosse, 1995; McLeod et al., 1996), and also need to be adequately trained mathematically in their content courses in a manner that is consistent with these reform movements.

This article will illustrate how a reform curriculum can be designed, implemented, evaluated, and departmentalized using the author's experiences reforming the core curriculum for undergraduate mathematics teaching majors. A brief review of the literature that discusses the research and educational trends that are motivating the current reform effort will be followed by an outline of the progression of mathematics reform within the Department of Mathematical Sciences at the University of Northern Colorado (UNC). Included in this discussion will be the basic assumptions and beliefs held by the author along with the grant activities that existed within the department that motivated the decision to reform the curriculum presented to undergraduate students passing through the mathematics teaching major. Lastly, an example of how this reform effort has affected specific students will be documented by examining their comments made during in-depth interviews discussing their experiences in the reformed core courses.

A Brief Accounting of the Origins of Reform in Mathematics

Cognitive Science

Cognitive research has shown that the process of learning is more complicated than was generally believed (Schoenfeld, 1987a; Schoenfeld, 1987b; Dubinsky, 1991; Tall, 1991). Silver (1987) describes the basics of cognitive theory and its impact on mathematics education. He states that students need to see mathematical thinking modeled in the classroom and that the "modeling needs to focus not only on *what* is being done but also on *why* the choice was made" (p. 56). Further he suggests that "In designing mathematics curricula . . . we need to be mindful that students will integrate their experience with the activity, unit, or course that we are preparing with their prior experiences to form or to modify attitudes toward and beliefs about mathematics and mathematical problem solving" (p. 57). Dubinsky and Lewin (1986) maintain that even "the best and most dynamic instruction will fail if it does not take into account the cognitive structures — both those possessed and those that must be acquired — of the knower, as well as the process (reflective abstraction) by which these constructions take place." Thus, as a consequence of the influence of cognitive science research on what it means to learn mathematics, the ways in which mathematics is taught must change. Instructors need to design instructional activities that will engage and then support their students' learning processes. Examples of effective instructional techniques are the use of a constructivist approach to teaching, the use of cooperative learning groups in the classroom, and the use of computer/calculator technologies to engage students in exploration activities.

The recognition of students' prior experiences (or lack thereof) and the need for the student to construct their own mathematical understanding illustrates how constructivism has changed mathematics teaching. When contrasting behaviorism with constructivism Tall (1991) stated that the "Constructivist psychology, on the other hand, attempts to discuss how mental ideas are created in the mind of each individual." A teacher-centered learning theory such as behaviorism or associationism viewed the teacher as the source of information and the student as the receiver with little attention given to the student's current understanding of the topic. However, constructivism supports an environment that is learner-centered where the student is actively constructing his or her own understanding using their prior knowledge as the foundation and building blocks for new concepts. Such a shift in learning theories naturally requires a shift in the pedagogical methods utilized and the goals established by the teacher. With respect to the undergraduate mathematics classroom, Steen (1989) stated, "The instructor's central goal should be to teach students how to learn mathematics." He continues by stating,

Students should gain an ability to read and learn mathematics on their own. Such maturity is as much a function of *how* mathematics is learned as *what* is learned . . . Students should be led to discover mathematics for themselves, rather than merely being presented with the results of concise, polished theories.

To make sense of new material, students must recall the appropriate background knowledge needed and then either "piece together" (reconstructing) or "pick apart" (deconstructing) the new material on their own. The tacit belief of the instructor is that students will have sufficiently digested previous concepts, attempted the homework problems, and will ask questions if they need further explanation. When left to their own devices, students who lack either the mathematical maturity or motivation may never successfully digest new material presented in a lecture format. Students that cannot or do not master new material quickly and independently become trapped in a pattern of failure from which they cannot extricate themselves. One means for avoiding this trap and establishing a student-centered learning environment is the use of cooperative learning groups. By allowing students to interact with one another in the classroom for extended time periods on meaningful tasks, the students have time and the opportunity to use their collective knowledge and reasoning abilities which become the building blocks for their learning. Such groups can be used to access previous knowledge upon which new mathematical ideas can be built, to summarize and distill concepts, or to discover mathematical connections. This also gives the instructor an opportunity to determine students' mathematical maturity and assess whether or not they have the abilities to "unpack" material presented in a lecture format. In their literature review of research in undergraduate mathematics education, Becker and Pence (1994) list the benefits of using cooperative learning groups as "active student involvement, opportunity to communicate mathematically; a relaxed, informal classroom atmosphere; freedom to ask questions; a closer student-teacher relationship; high level of student interest; more positive students attitudes; and opportunity for students to pursue challenging mathematical situations." Numerous studies have demonstrated that these potential benefits from using cooperative learning groups are a reality (DePree, 1998; Kaufman et al., 1997; Hinzman, 1997; Leikin & Zaslavsky, 1997; Brush, 1997; Nichols & Hall, 1995; Rosenthal, 1995; O'Malley & Scanlon, 1990). Such positive experiences influence students significantly in their ability to do mathematics and their desire to continue to study mathematics when it is no longer required by their programs of study (McLeod, 1992; Garofalo, 1989; Meyers, 1993; Duren and Cherrington, 1992).

The use of technology to enhance mathematics learning has also changed the practice of instruction. In his preface to the MAA Note *Reshaping College Mathematics*, Steen (1989) states that "students should

make full use of calculators and computers in all mathematics courses."
Access to cheap and powerful technology allows both the teacher and
student to hand over more mundane and repetitive work to a machine or
program. Rather than viewing such "cognitive technologies" as mere
alleviators of the mundane, Pea (1987) suggests that they act as amplifiers
that influence not only *what* we teach using technologies but also *how* we
teach using these technologies. He states that "The dynamic and interactive
media provided by computer software make gaining an intuitive understanding (traditionally the province of the professional mathematician) of
the interrelationships among graphic, equational, and pictorial representations more accessible to the software user. Doors to mathematical thinking
are opened, and more people may wander in."

Pea's comments refer to the use of any cognitive technology. Dubinsky
(1991), for example, focuses on using a computer programming language,
ISETL, to enhance student learning. He contends that requiring students to
write their own computer programs (as opposed to just utilizing a program)
to construct mathematical objects forces students to mentally construct the
mathematical object as well. He states "activities with computers are a
major source of student experiences that are very helpful in fostering
reflective abstractions" (p. 123). Research supports the proposition that the
use of technology at the collegiate level improves student achievement and
encourages positive attitudes (Peck et al., 1994; Quesada and Maxwell,
1994; Guckin and Morrison, 1991; Stiff et al., 1992). Sommerfeld (1995)
stated that the Committee for Economic Development report found that
computers improved students' abilities to visualize models and concepts
and to analyze data; Connell (1998) warned, however, that to improve
student learning technology's use must be aligned with constructivist
philosophies.

Societal Pressures

National reports such as *Renewing U.S. Mathematics: Critical
Resource for the Future, Everybody Counts* and *A Nation at Risk* have also
influenced the reform movement. These reports highlighted the state of
mathematics learning in the United States and the need for increased
national mathematical ability in a highly competitive and technologically
rich global market. These and similar reports provided the impetus for the
development of the NCTM Standards. McLeod et al. (1996) provides an
excellent timeline of the national and international forces that influenced

the development of the NCTM Standards. At the undergraduate level, the MAA Committee on the Undergraduate Program in Mathematics (CUPM) has attempted to "provide coherence to the mathematics major by monitoring practice, advocating goals, and suggesting model curricula" (Steen, 1989).

Today's job market demands workers who are mathematically and technologically literate. The National Research Council's 1991 *Moving Beyond Myths* report stated that "Prosperity in today's global economy depends on scientific and technological strength, which in turn is built on the foundation of mathematics education" (p.1). In response to this technology-driven market place, the number of calculus and post-calculus mathematics courses required by university majors has increased. Duren (1994) states that "advanced mathematics has become more necessary than ever before in old and new fields of science, business, and technology that are vital to America's welfare." Mathematics courses such as abstract algebra in which only mathematics majors used to enroll now support diverse fields. Consequently, not only has the audience for advanced mathematical courses changed but also, the qualifications, mathematical backgrounds and interest level of these students has changed (Becker and Pence, 1994; Steen, 1989). Reform has become increasingly relevant in addressing these changes.

A Roadmap of Departmental Reform in Retrospect

Assumptions/Beliefs upon Which the Reform Effort Was Built

Reform in teaching methods and curricula is pervasive in the Department of Mathematical Sciences at UNC due to the strong contingency of mathematics educators housed in the department, the interest of mathematicians in pedagogy and the influence of the teaching assistants in the Educational Mathematics Ph.D. program. Research is on going in a wide range of courses from service courses (e.g. Mathematics for Liberal Arts and business calculus) to upper-division courses such as abstract algebra. A core belief held by the faculty was that change prompted by research and reform was not only natural, but also desired.

The role of UNC as the premier teacher education institution in Colorado and the influence of the Ph.D. program resulted in the instructors of content courses for pre-service teachers paying special attention to instruction and assessment so as to be consistent with the goals of the NCTM Standards. This change was also a result of the belief that teachers will teach mathematics as they have been taught mathematics and that a handful of mathematics methods courses would not effectively change students' attitudes about mathematics or the teaching and learning of mathematics. Such a change would need to come from learning *mathematics* in an environment that models teaching methods consistent with the learning theories espoused by current reform efforts such as the Standards. The instructors in the department believed that to effect a lasting impact on a student's learning behaviors, beliefs/attitudes about mathematics, and eventually their teaching behaviors exposure to quality teaching and meaningful content must be *prolonged* and *consistent* among content courses. They also felt that in order to make learning "stick," content/curricular themes must be repeated and broadened as students progressed through their undergraduate program.

A partnership with local school districts, supported by a US West fellowship, allowed the faculty to be influenced by veteran classroom teachers, which thus influenced the mathematics instruction at the pre-collegiate level. A master teacher from one of the school districts joined the department for a year as an instructor and teaching resource.

Model of reform at UNC

In Colorado, all teaching candidates (elementary, middle and secondary) must declare a major in a specific content area, such as mathematics, in addition to a minor in education. This can result in elementary school teachers having considerable depth in their mathematical background because of the number of content courses required. All mathematics majors are required to take two semesters of calculus, linear algebra, discrete mathematics and informal geometry. An additional third semester of calculus, abstract algebra, modern geometry, statistics, and probability are required of secondary education mathematics majors. The number of content courses required of all mathematics teaching majors provided the faculty with multiple opportunities to impact students' backgrounds in mathematics, their attitudes and beliefs about teaching.

Those methods shown to be effective in improving student attitudes and learning either via research in the literature or from on-going research within the department were implemented in the classroom. Specifically the use of a constructivist pedagogy, reading, writing, verbalizing, appropriate integration of technology, cooperative groups and striving for conceptualization rather than just proceduralization of the mathematics were incorporated into many of the core content courses offered. In addition to using consistent pedagogy in these courses, the faculty tried to draw explicit connections between topics within and between courses in the curriculum. A single topic like differentiation can be discussed in multiple courses using different viewpoints to encourage students to broaden their understanding of the concept of the derivative of a function. A discussion of how this can be accomplished appears later in the article.

The Road of Reform Travelled

Description of a Trio of Linked Reformed Courses

Linear algebra, discrete mathematics, and abstract algebra were reformed and interrelated as part of this project. The content contained in the three courses was fairly traditional. Linear algebra began with Gaussian elimination and proceeded through vector spaces. Discrete mathematics focused primarily on combinatorial mathematics to include data collection, patterns, recursions, equivalence relations, etc. as opposed to discrete structures and graph theory. Abstract algebra ran the gamut from mappings to quotient groups. In addition to wanting students to learn "good mathematics," the goals of this reform effort were to design a curriculum that (1) would inspire students to think abstractly and to appreciate the need for abstraction, (2) fostered independence in their learning of abstract mathematics, (3) enabled understanding and valuing the need for mathematical proof, and (4) facilitated communication of their understanding to other individuals.

A "holistic spray" (Asiala et al., 1996) of a variety of pedagogical approaches was utilized in teaching these courses. Working in cooperative groups, students were provided with activities designed to summon up essential background knowledge, to provoke conflict within their concept image, to challenge them to construct mathematical concepts or procedures, to solve problems, and to communicate their understanding of concepts to other members of the class. The students were taught how to read and summarize mathematical prose. After reading the text, they wrote summaries, generated examples, and listed two types of questions —

review questions they could answer and questions that articulated concepts that were still unclear from their initial reading. In linear algebra graphing calculators were used to explore matrices and their properties; homegrown computer programs were used to investigate linear transformations. The discrete mathematics course incorporated the use of graphics calculators (specifically TI-92s) and spreadsheets to explore data collection, recursions, polynomial curve fitting, etc. In abstract algebra the computer program *Exploring Small Groups* was used in conjunction with labs written by one of the instructors. Labs were designed to reinforce concepts and to investigate new material such as normality, index, and quotient groups. Most importantly, the students were explicitly told and reminded they were expected to be active participants in the classroom and responsible for their own learning.

Instructional Innovations Employed

Several of the courses were team-taught. Throughout the semester a variety of instructional innovations designed to encourage students to talk to one another and to the instructors were developed. Extended study sessions for all three classes simultaneously were offered. Each course was assigned a one-hour time during a three-hour study session in which they were allowed to ask questions for presentation on the board, reserving the balance of the time for individual or small group help. These sessions allowed interaction between students from different courses prompting comments like "you mean you use onto in your course too?" and abstract algebra students reviewed content they learned in linear algebra and linear algebra students saw a "preview of coming attractions" of their future courses. The study sessions were scheduled in the early evening hours to maximize student attendance and to meet the needs of nontraditional students. Prior to major exams, a Jeopardy-like game involving group verbal responses served as a review and as a "free" assessment tool. Students commented "I found in listening to responses that I was just as good as the rest of the class."

Another innovation used in abstract algebra was to take the chalkboard and overhead away from the instructors (and students). Students were asked to talk about normality and to describe quotient groups. During these "chalkless" talks subtle misconceptions on the part of the students were revealed; these were addressed in the following class period.

In abstract algebra, index cards containing a statement or small theorem to be proven were circulated at the beginning of the semester so that each student could select one to present. Each student prepared a proof, privately presented it to one of the instructors, and after several iterations was allowed to present it to the class and to respond to instructor and student questions. This process provided each student with the opportunity to spend extended quality time with the instructors and to appreciate the high level of preparedness required for understanding and ultimately presenting a proof. This also encouraged students to take ownership of their mathematical statement and to become an expert on that particular topic. The cooperative groups in abstract algebra were given a similar opportunity. During the second class period, each table of four was given eight elements and a binary operation, and was asked to develop the operation tables for one of the five non-isomorphic groups of order eight. Each cooperative group gave a presentation on their group, and once properly scrutinized and revised, copies were distributed to the rest of the class. These operation tables played a crucial role in learning about order, centers, and isomorphism.

Spiraling Topics Through the Sequence

As instructors and as mathematicians, professors hope that students will see and appreciate the connections between courses and concepts. The strength and depth of these connections can serve as a means for anchoring a student's understanding and enhancing their ability to recall that knowledge in problem solving situations. Students typically fail to make such connections on their own. This is partly due to their lack of mathematical maturity, but is also a consequence of their attitudes and beliefs about mathematics. Students' develop negative attitudes and beliefs, including the view that it is an unchanging, disconnected discipline, as a result of the curricula to which they are exposed and the continued use of an absorption model of teaching (Goldenberg, 1991; Hyde & Hyde, 1991; McLeod, 1992). Their belief that mathematics is a disconnected field precludes their attempt to connect content from one course to that of another. To overcome this blockade, Silver (1987) suggests making the "hidden curriculum" transparent. He concludes that explicitly teaching students the connections and habits of mind we want them to have will improve their learning and attitudes in mathematics.

In addition to the common threads of functions, proofs, logic, data collection, and the use of counterexample, there are multitudes of opportunities to link this trio of courses to each other and to others such as calculus and modern geometry. Discussions of substructures can link the concepts of subset, subspace, and subgroup. Students can learn proof by mathematical induction by practicing on identities from calculus (derivative of x^n), geometry (number of diagonals in a convex n-gon), or trigonometry (DeMoivre's Theorem). Coset decomposition in group theory is foreshadowed first in set partitions and again in equivalence relations. Differencing sequences of numbers to fit a polynomial to data is a discrete version of differentiating. The characteristic root technique and iteration for solving recursions are reminiscent of linear combinations of vectors in a basis and composition of functions. Long division of polynomials can be linked to infinite series and Taylor polynomials from calculus connecting to the use of generating functions for solving recursions in discrete mathematics, which then connects to solving systems of linear equations using matrices from linear algebra. The next three examples will be explored in more depth.

Properties of real numbers with respect to addition and multiplication are already well formed by the end of first year algebra. However, the real numbers form a field and thus behave in an idealized manner satisfying properties that other sets of numbers do not. This may skew students' understanding of properties attributing them to the operation rather than the set of numbers. The first day in linear algebra and in abstract algebra the students brainstormed the properties of the real numbers and then began to investigate these properties as encountered in their courses. The linear algebra students considered matrix algebra (this could also be done when starting with vector spaces) and the abstract algebra students considered Cayley tables of groups and non-groups of order 4. The students found that in ninth grade algebra and when working with an abstract group you can cancel, but in linear algebra when working with matrices (and in discrete mathematics working with sets) you cannot. Students in both classes initially struggled to list all the properties of the real numbers under addition and multiplication. This information was crucial to their success and ability to understand basic content in the two courses. In a graduate abstract algebra course, the students were able to extend this activity by constructing the definition of a group from the examples/non-examples of groups given. Their definition did not require the elements to commute, underscoring in their mind the need for the definition of abelian.

A second example of connections that can be pointed out explicitly to students involves the nth roots of unity found by solving the equation $x^n - 1 = 0$. This problem encourages students to see connections between geometry, vectors, group theory, algebra, and long division. While working on this problem in abstract algebra students used concepts from linear algebra and transformational geometry to determine the roots. They could see using either vector addition or by changing the roots into polar form that this set of vectors was closed under the binary operation. In both cases, the students were able to review concepts from previous courses and improve their understanding of the old and new concepts. Also, a nice link to discrete mathematics and the finite geometric sum emerges.

$$1 + x + x^2 + x^3 + \ldots + x^{n-1}$$

Finally, proving identities involving the Fibonacci numbers provide a solid connection between linear algebra, discrete mathematics, number theory, and abstract algebra. In discrete mathematics students use their background in linear algebra and newly honed skills in proof by mathematical induction to prove the matrix equation

$$\begin{bmatrix} 0 & 1 \\ 1 & 1 \end{bmatrix}^n = \begin{bmatrix} F_{n-1} & F_n \\ F_n & F_{n+1} \end{bmatrix}$$

Combining this with the homomorphism property, $\det AB = (\det A)(\det B)$, from abstract algebra allows students to prove identities such as

$$F_{n-1}F_{n+1} - F_n^2 = (-1)^n.$$

Student Feedback on the Process

How Did Students Respond to the Reform?

Under a team teaching format, the main components of the reform implemented were: using constructivist teaching methods and cooperative groups; emphasizing the efficacy of reading, writing, and verbalization as learning tools; integrating technology into the instruction and learning of the material; and, focusing on conceptualization and explicitly linking concepts formed with other fields of mathematics. Additionally, the innovations adopted by the pair of instructors emphasized: modeling active learning through peer interaction within the classroom, creating a student-teacher partnership in the learning process, and encouraging student ownership of the mathematics learned. The effect these changes had on students will be discussed in light of the following two quotes. While these quotes were taken from only two of the student interviews, they are representative of the comments contained in all twelve of the student interviews.

> The smaller groups seem to work better, at first at least, because when you are talking to the whole classroom sometimes it can be embarrassing so then if there are four or five people in a group then they get to know each other and they start to work well together and work more like a team. It works well I think because if one person is stuck at a certain spot the next person may not be. When one person gets stuck, there are two or three or four other people to figure out where to go from here (Student A, linear algebra).

> The reason I've learned as much as I have is because I come in and I'm the teacher of the group. Even when we got in a study group they all asked me, "How do you do this?" And all of a sudden I had to think about it and since I was putting it in my own words and how I would do it, all of a sudden on the test it was a piece of cake. That's how I attribute my success (Student B, abstract algebra).

Both of these students became very comfortable working in their cooperative groups. The power of working in teams came from the collective resources available to bring to bear on a given problem. As pointed out by Student A, where an individual may have gotten stuck on a problem, the group was able to continue to work because of the additional resources. The groups were able to dispel many simple misunderstandings, to fill in needed background material and to provide multiple perspectives. When the entire group lacked the skills needed to proceed, the fears related to asking questions of the instructors in front of the class were minimized because there were four or five individuals asking the question, not just one. They also used group members as sounding boards and as a means for rehearsing their verbalization of their understanding of concepts. As illustrated by Student B's comment, putting concepts into her own words and describing to others how she would work through a problem solidified her understanding of the material. Her experience shows how the use of constructivist teaching techniques, working in cooperative learning groups, and the focus on verbalizing their understanding can positively affect not only how well a student learns concepts, but also how well they can demonstrate their understanding in an assessment setting. This student's experiences are consistent with Vygotsky's (1978) research in which he suggested the self-talk used to monitor and direct thinking is usually internalized from a student's interactions with the instructor or with other students when explaining or justifying his or her reasoning.

Using computers or calculators as a means for generating data and exploring "what if" questions integrated the use of technology into the courses in a meaningful way. The inclusion of these tools encouraged the students to work hard to obtain a good conceptual understanding of the material rather than been sated by a working procedural knowledge. Students C and D describe their interaction with technology in the course and its impact on their learning.

> The biggest thing for me I guess, the most beneficial part was the quotient one. And it goes back to I could do it but I didn't understand what it was all about. To see an elementary picture, to see the blocks are just groups of elements. That really helped. I guess it depends on what kind of learning you're after. I think the colored blocks where each little block was a coset, as elementary as it sounds, the colored blocks helped a lot for visualizing. 'Cause then when you go back and you're grinding out a quotient group you go back and you visualize, "oh it's just that green block" (Student C, abstract algebra).

> I really appreciated the straightforward manner the labs are written. It allowed me to focus on what I was trying to learn rather than on which key I supposed to remember to push. It generated so much data, so quickly. I wouldn't have been able to do that by hand. I was able to concentrate on figuring out what was really going on in the problem (Student D, discrete mathematics).

Of particular importance to whether or not students would gain from the introduction of a technology tool was how seamless and meaningful the integration of that tool was in the learning process. This observation is consistent with the department's experience with the use of a Mathematica lab in the calculus sequence. Students became frustrated with the apparent lack of connection between what they were doing in the lab and what they were learning in the course. Their learning was also hampered by the typical difficulties a novice may experience when working with a programming CAS like Mathematica.

Students also recognized the need for connections to be made between and within courses. One linear algebra student in particular referred to this need with respect to his future teaching.

> Part of the problem with how much theory we learn is that it's so long ago that we practiced basic calculus when you get back to teaching high school we have to go through and almost rework it (all), because I haven't done it in four years. Like everything else, if you haven't done it in four years, you are not going to remember it.

The need for seeing such connections and retaining concepts from previous courses is becoming increasingly important in the face of state testing. These innovations also strengthened the student-teacher bond. In abstract algebra 24 of 25 successfully completed the course; the 12 female students were particularly successful. Final exam results and student attitudes were superior to those experienced by one of the instructors when teaching a comparable course a decade earlier.

Long Term Impact on Student Behavior

The long-term impact on the students manifested itself in their behavior in subsequent courses, in their student teaching internship, and in their attitude about mathematics. The students formed a network that they could turn to in their current course and subsequent courses. Other department faculty reported that these students displayed a natural proclivity to work in teams and to actively engage in classroom discussions, including anticipating key concepts in their conjectures.

Seeing the teaching strategies that they would be expected to implement during their teaching career modeled in their content classes encouraged these students to utilize these same methods during their student teaching internship. The student teacher supervisors from the College of Education reported back to the department that the students were "increasingly using cooperative learning groups to engage students in the learning process." There was also a noticeable increase in the use of technology even when the home teacher did not actively promote the use of calculators in the classroom.

The students showed an increased awareness of the connections between fields of mathematics, the need to work hard to gain understanding, and a positive attitude toward the need for mathematical abstraction. One student stated that the impact her experiences had on her was "It just made me work harder I think (in modern geometry). I think it (linear algebra) made it easier for me to go on because of the habits I learned." This same student also spoke of connections between linear transformations from linear algebra and the topics she learned in modern geometry which was not a course involved in this reform effort. A student from abstract algebra who had gone through all three reformed courses described what abstract algebra was like for him. His description is atypical of how students generally feel when leaving their first course in group theory; however, his comments were typical of the six abstract algebra students interviewed.

> You take all the tools you have had in other courses and apply them to groups. You make connections between subjects: matrices, permutations, etc. It helps clarify subjects that we have hit in the class, like matrix multiplication, one-to-one and onto. It's frustrating, but fun. I like being stumped. I don't like being force-fed. I like big connections. We need to understand the theme, and see why it is important.

Lessons Learned along the Way

Extending Reform Beyond the Course and Individual Instructor

This project began as a small part of a statewide reform effort supported by an NSF grant. Linear algebra was the first course that was reformed; that effort became the subject of a Ph.D. study. The reform team working on linear algebra was composed of two mathematicians and a student in the educational mathematics Ph.D. program. This composition allowed the three to monitor both the mathematical content included in the course and the pedagogical methods. It also promoted a level of acceptance within the two facets of the department (mathematics education and mathematics). The success of the linear algebra reform effort (Mingus, 1994) encouraged other members of the department to develop similar reform efforts in other content courses such as modern geometry.

While co-teaching was utilized in linear algebra at times as part of the reform, it was not an essential component. As the reform effort was extended to include first discrete mathematics and then abstract algebra, the use of team teaching became an integral part. Once again, the team was composed of a mathematician and a mathematics educator. Not only did it provide the instructors with mutual encouragement to continue down a path that is often fraught with difficulty and criticism, but this dual perspective also proved to be greatly beneficial for the students. Team teaching and planning also provided the two instructors with a broader perspective when determining which topics to include, which instructional methods to use and how assessment could be conducted.

Renown originator of quantum theory Max Planck pessimistically said that "A new scientific truth does not triumph by convincing its opponents and making them see the light, but rather because its opponents eventually die out, and a new generation grows up that is familiar with it." Planck's statement seems to apply to the acceptance of reform mathematics curriculum and pedagogy over traditional. However, the experiences at this university suggest that encouraging other faculty members to become familiar with reform efforts and providing them with evidence of the potential impact on student learning and attitudes can persuade others to continue with the changes initiated. Through national exposure at conferences and peer-reviewed articles in reputable journals additional credibility of the reform implemented can be garnered, thus convincing other department members "to buy into" the new curriculum.

Research efforts in studying the effects of reformed curriculum provide extra incentive (i.e. John Henry effect) on the part of department members to continue to work hard during the implementation phase. Also the results of the research can provide crucial support for the efficacy of the efforts and encourage others in the department to push for making the reformed methods the accepted way to teach the core courses.

Caveats in the Reform Process

However pervasive reform may become in mathematics education, the university setting may be the last bastion of academic freedom. As a result of this independence, conflict among department members can arise with respect to which or whether or not reform curricula and/or teaching methods should be used. How conflicts between faculty are resolved can influence the effectiveness of specific reforms within the department and whether or not faculty are willing to risk reforming courses at all.

Since reforming any course is time and energy intensive, developing course materials for broad use that document the reform helps "departmentalize" some reform methods. Reform is iterative in nature and so this record will give the department and individual faculty members a source of information about the reasons motivating the reform and documentation of what worked and what did not and why. Also, maintaining a record of the reforms attempted will alleviate the burden of developing supplementary course materials on those faculty interested in utilizing reformed curriculum but who might either be "riding the fence" as to the reform's value or feel uncertain about his/her ability to develop appropriate learning materials.

While "buy in" to reform is important, there is a distinct danger in departments or individual instructors becoming devoutly dedicated to a particular set of reforms. Such dedication is essential at the outset of a reform effort in order for time, energy and resources to be devoted to its development and use. Instructors can become so personally attached to "their method" of reform that any suggested change could be interpreted as a personal attack on their ability to teach. However, if faculty cease to question the efficacy of the reform it will be difficult to see or anticipate its weaknesses or failures. Departments may also be reticent to abandon a given reform effort because of the investment of considerable resources in the form of equipment or training of faculty members. A certain level of resistance to abandoning reform efforts undertaken in general education or service courses is also natural. Departmental reputation may be viewed as "at stake" if the decision is made to change the curriculum again.

In Closing

Successful reform efforts tend to be difficult to replicate. While some aspects of a given reform such as computer explorations can be extracted and exported from classroom to classroom, others are elusive. Reform occurs not just within a curriculum, but within human beings. The students in the classroom, the teacher's motivation and enthusiasm for utilizing the curriculum, the attitudes in the department and at administrative levels about reform in general, the resources committed and many other factors can determine success or failure. Some lessons that were learned in the process of this reform effort were: (1) start with a small project and do not let it grow too quickly, (2) be vigilant in determining what works and what does not work; (3) find external sources of support; (4) seek out allies within the department and from other departments that may use the course as a prerequisite; and (5) listen very carefully to the harshest critics and question the most ardent supporters.

References

Asiala, M; Brown, A.; Devries, D.: Dubinsky, E.; Matthews, D.; and Thomas, K. (1996). A framework for research and curriculum development in undergraduate mathematics education. In *Research In Collegiate Mathematics Education* (vol. II), eds. James J. Kaput, Alan H. Schoenfeld, and Ed Dubinsky, 1-32. Providence, RI: American Mathematical Society.

Becker J. R. and Pence, B. J. (1994). The teaching and learning of college mathematics: current status and future directions. In *Research Issues in Undergraduate Mathematics Learning: Preliminary Analyses and Results* (MAA Notes 33), eds. James J. Kaput and Ed Dubinsky, 5-16. Washington, DC: Mathematical Association of America.

Bosse, M. J. (1995). The NCTM standards in light of the new math movement: A warning! *Journal of Mathematical Behavior*, 14, 171-201.

Brush, T. A. (1997). The effects on student achievement and attitudes when using integrated learning systems with cooperative pairs. *Educational Technology Research and Development*, 45(1), 51-64.

Connell, M. L. (1998). Technology in constructivist mathematics classrooms. *The Journal of Computers in Mathematics and Science Teaching*, 17(4), 311-38.

DePree, J. (1998). Small-group instruction: Impact on basic algebra students. *Journal of Developmental Education*, 22(1), 2-6.

Dubinsky, E. (1991). Reflective abstraction in advanced mathematical thinking. In *Advanced Mathematical Thinking*, ed. David A. Tall, 95-126. Dordrecht, The Netherlands: Kluwer Academic Publishers.

Dubinsky, E. and Lewin, P. (1986). Reflective abstraction and mathematics education: The genetic decomposition of induction and compactness. *The Journal of Mathematical Behavior*

Duren, W. L. Jr. (1994). The most urgent problem for the mathematics profession. *Notices of the American Mathematical Society*, 41(6), 582-586

Duren, P. E. and Cherrington, A. (1992). The effects of cooperative group work versus independent practice on the learning of some problem-solving strategies. *School Science and Mathematics*, 92(2), 80-83.

Ferrini-Mundy, J. and Johnson, L. (1997). Highlights and implications. In *The Recognizing and Recording Reform in Mathematics Education Project: Insights, Issues and Implications* (Journal for Research in Mathematics Education, Monograph Number 8), eds. Joan Ferrini-Mundy and Thomas Schram.

Garofalo, J. (1989). Beliefs and their influence on mathematical performance. *Mathematics Teacher*, 502-505.

Goldenburg, P. E. (1991). Seeing beauty in mathematics: Using fractal geometry to build a spirit of mathematical inquiry. In *Visualization in Teaching and Learning Mathematics*, eds. W. Zimmerman and S. Cunningham, 39-66. N. P.: Mathematical Association of America.

Guckin, A. M. and Morrison, D. (1991). Math*logo: A project to develop proportional reasoning in college freshman. *School Science and Mathematics*, 91(2), 77-81.

Hyde, A. A. and Hyde, P. R. (1991). *Mathwise: Teaching mathematical thinking and problem solving*. Portsmount: Heinemann Educational Books, Inc.

Hinzman, K. (1997). *Use of Manipulatives in Mathematics at the Middle School Level and Their Effects on Students' Grades and Attitudes*. An unpublished master's thesis, Salem-Teikyo University.

Kaufman, D., Sutow, E. and Dunn, K. (1997). Three approaches to cooperative learning in higher education. *Canadian Journal of Higher Education*, 27(2-3), 37-66.

Leikin, R. and Zaslavsky, O. (1997). Facilitating student interactions in mathematics in a cooperative learning setting. *Journal for Research in Mathematics Education*, 28(3), 331-54.

Mathematical Sciences Education Board. (1989). *Everybody Counts*. Washington, DC: National Academy Press.

McLeod, D. B. (1992). Research on affect in mathematics education: A reconceptualization. In *Handbook of Research on Mathematics Teaching and Learning*, ed. D. A. Grouws, 575-596. Reston, VA: National Council of Teachers of Mathematics.

McLeod, D. B., Stake, R. E., Schappelle, B. P., Mellissinos, M., and Gierl, M. J. (1996). Setting the Standards: NCTM's role in the reform of mathematics education. In *Bold Ventures: Case Studies of U.S. Innovations in Mathematics Education*, vol. 3, ed. Senta A. Raizen and Edward D. Britton, 13-132. Dordrecht, The Netherlands: Kluwer Academic Publishers.

Meyers, N. C. (1993). Cooperation in calculus. *PRIMUS*, 3(1), 47-52.

Mingus, T. Y. (1994). *A Qualitative and Quantitative Study Examining the Effect a Conceptual, Constructivist Approach to Teaching Linear Algebra has on Student Attitudes and Beliefs about Mathematics*. An unpublished dissertation.

National Commission on Excellence in Education. (1983). *A nation at risk: The imperative for education reform*. Washington, DC: US Government Printing Office.

National Council of Teachers of Mathematics. (1980). *An Agenda For Action: Recommendations For School Mathematics Of The 1980s*. Reston, VA.

National Council of Teachers of Mathematics. (1989). *Curriculum And Evaluations Standards For School Mathematics*. Reston, VA.

National Research Council. (1991). *Moving Beyond Myths*. Washington, DC.

Nichols, J. D. and Hall, N. (1995). The effects of cooperative learning on students achievement and motivation in a high school geometry class. Paper presented at the 1995 Annual Meeting of the American Educational Research Association.

O'Malley, C. E. and Scanlon, E. (1990). Computer-supported collaborative learning: Problem solving and distance education. *Computers and Education*, 15(1-3), 127-36.

Pea, R. D. (1987) Cognitive technologies for mathematics education. In *Cognitive Science and Mathematics Education*, ed. A. H. Schoenfeld, 1-32. Hillsdale, NJ: Lawrence Erlbaum Associates, Inc.

Peck, R. Jean, B. and Shaw, N. (1994). A statistical analysis on the effectiveness of using a computer algebra system in a developmental algebra course. (Reprint)

Quesada, A. R. and Maxwell, M. E. (1994). The effects of using graphing calculators to enhance college students' performance in precalculus. *Educational Studies in Mathematics*, 27, 205-215.

Renewing U.S. Mathematics: Critical Resources for the Future. (1984) *Report of the Ad Hoc Committee on Resources for the Mathematical Sciences.* Washington, DC: National Science Foundation.

Rosenthal, J. S. (1995). Active learning strategies in advanced mathematics classes. *Studies in Higher Education*, 20(2), 223-28.

Schoenfeld, A. (1987a). Cognitive science and mathematics education: An overview. In *Cognitive Science and Mathematics Education*, ed. A. H. Schoenfeld, 1-32. Hillsdale, NJ: Lawrence Erlbaum Associates, Inc.

Schoenfeld, A. (1987b). What's all the fuss about metacognition? In *Cognitive Science and Mathematics Education*, ed. A. H. Schoenfeld, 1-32. Hillsdale, NJ: Lawrence Erlbaum Associates, Inc.

Silver, E. A. (1987). Foundations of cognitive theory and research for mathematics problem-solving. In *Cognitive Science and Mathematics Education*, ed. A. H. Schoenfeld, 1-32. Hillsdale, NJ: Lawrence Erlbaum Associates, Inc.

Sommerfeld, M. (1995). Report links access to technology to math, science reform. *Education Week*, 15.

Steen, L. A. (1989). Reshaping college mathematics: A project of the Committee on the Undergraduate program in mathematics. *MAA Notes Number 13*. Washington, DC: Mathematical Association of America.

Stiff, L. V., McCollum, M. and Johnson, J. (1992). Using symbolic calculators in a constructivist approach to teaching mathematics of finance. *Journal Of Computers In Mathematics And Science Teaching*, 11, 75-84.

Tall, D. (1991). The psychology of advanced mathematical thinking. In *Advanced Mathematical Thinking*, ed. David A. Tall, 95-126. Dordrecht, The Netherlands: Kluwer Academic Publishers.

Vygotsky, L. S. (1978). *Mind In Society: The Development Of Higher Psychological Processes*. Cambridge, MA: Harvard University Press.

3

DELINEATING THE RELATIONSHIP BETWEEN PROBLEM-SOLVING BEHAVIORS AND ACHIEVEMENT OF STUDENTS IN COOPERATIVE-LEARNING GROUPS

Christine Ebert
Patrick Mwerinde

Introduction

This study examined the problem-solving behaviors, strategies, and achievement of college students enrolled in a one-semester College Algebra and Statistics course, with respect to the content area of quantitative literacy, connections between algebraic and graphical representations, and mathematical modeling. Four instructional units of this course were chosen — two in which the students were assigned to cooperative learning groups and two in which the students worked independently. The findings suggest that students who work in cooperative-learning groups clearly exhibit important problem-solving behaviors such as persistence and a willingness to explore alternative solutions. In particular, the students in the cooperative learning groups engaged in the type of mathematical discourse that would allow them to form connections between graphical and algebraic representations. However, both groups of students still experienced difficulty explicating the connections between mathematical actions and/or processes and mathematical concepts.

Conceptual Framework

Cooperative learning strategies have been credited with the promotion of critical thinking, higher-level thinking, and improved problem-solving ability of students. Current research that examines behaviors that occur during group problem-solving sessions seem to indicate that groups engage in behaviors that are similar to those exhibited by expert mathematicians when they solve problems (Arts & Newman, 1990; Schoenfeld, 1987): that is, they engage in monitoring their own thoughts, the thoughts of their peers, and the status of the problem-solving process. Researchers who have studied cooperative learning at the college level generally have found that students learn just as well as in more traditional classes and often develop improved attitudes toward each other and toward mathematics (Dees, 1991; Slavin, 1995; Brechting & Hirsh, 1977; Chang, 1977; Davidson, 1971; Olsen, 1973; Shaughnessy, 1977; Treadway, 1983). Although it is not clear which components of cooperative learning are responsible for improvements in higher-level thinking, attempts have been made to identify the contributing factors. One conjecture is that dealing with controversy may be such a factor. Smith, Johnson, and Johnson (1981) report on a study in which they suggest that higher results with respect to achievement and retention of information for the students in the "controversy group" may be attributed to the "cognitive rehearsal of their position and their attempts to understand their opponents position" (Smith, Johnson, & Johnson, 1981, p.652). This finding suggests that the process of engaging in discourse with respect to explaining and defending one's position, i.e., mathematical solution, enables the individual, with respect to mathematical knowledge, to construct more connections and strengthen the connections that exist. In fact, this type of discourse also describes quite well the behaviors exhibited by experienced mathematicians as they collaborate on the solution of a problem. Thus, the consonance of these findings suggest that the influence and role of discourse must also be considered throughout an examination of the relationships between problem-solving behaviors and achievement. In this study, we not only examined the problem-solving behaviors, strategies, and achievement of college students assigned to cooperative learning groups, but designed problem-solving experiences that would facilitate the discourse. These experiences were also consonant with the course curriculum and focused on explicating the connections between mathematical actions and/or processes and the underlying mathematical concepts.

Design of the Experiment

The experiment was designed to investigate the problem-solving behaviors, strategies and achievement of 108 university students enrolled in a College Algebra and Statistics course. The course consisted of 3 class meetings per week with 2 lectures and 1 lab or workshop. Six workshops per semester were held in the microcomputer lab. Two experimental groups and two control groups were randomly selected. On the first day of class an attitude survey and a pre-test of algebraic ability were administered to both the experimental and the control groups. Based on their performance on the pre-test of algebraic ability, students in the experimental group were assigned to cooperative learning groups. Each group consisted of 4 students (1 high score, 2 middle scores, 1 low score). In the control sections, students were encouraged to work with other students on the various activities and labs but were not specifically assigned to groups.

Throughout the semester, problem-solving behaviors, strategies, and achievement were assessed through four tasks which focused on the connections between mathematical actions and processes. The following list provides a brief description of the tasks that defined the study:

- Pretest of Algebraic Ability

  10 items (topics ranged from using percents to evaluating a logarithm and interpreting a graph)

- Pre-Attitude Scale

  Included questions to determine demographic data such as the number of years of university mathematics.

- Post-Attitude Scale

  Included additional questions about cooperative learning groups.

• Four Instructional Tasks

Focused on the connections between mathematical actions and processes and the mathematical concepts.

Task I: Regular classroom setting quantitative literacy.

Task II: Microcomputer lab connections between algebraic and graphical representations in the context of linear depreciation.

Task III: Microcomputer lab "Best model" from a data set.

Task IV: Regular classroom setting exponential modeling.

Conduct of the Study

Once the design of the study was determined and the various tasks such as the pretest, attitude scale, and instructional tasks were constructed, then an audio and video taping schedule was set up to collect the data. Each of the four tasks were videotaped (some problem-solving activities were recorded for each group) and audiotaped (the problem-solving discourse was recorded for each cooperative learning group within both experimental sections). In addition, the initial and final problem-solving sessions devoted to quantitative literacy and exponential modeling were also videotaped for the control sections. The written work that accompanied each of these tasks was also analyzed with respect to the students' ability to explicate the connections between actions and/or processes and mathematical concepts. At the end of the semester, the attitude scale with some additional open-ended questions concerning cooperative learning groups was again administered to all of the students.

Results

In order to determine between-group similarities and differences, the results of the MSAT, the pre-test of algebraic ability, and the questions concerning the number of years of high school mathematics and their previous university mathematics history were analyzed. Seventy-eight students comprised the experimental group and thirty students made up the control group. The results of these analyses are summarized in the following table.

Table 1: Baseline Data

	MSAT		P-test Score Mean n = 10	Years of High School Math Mean	University Math History		
Group	Mean	Std. Dev.			1st Class	Remedial	Na
Ctrl. n = 30	470	70.36	4.65	3.62	53%	37%	10%
Exp. n = 78	485.2	88.25	5.30	3.93	64%	28%	8%

These results indicate that all of the baseline assessments with respect to mathematical ability and preparation were consistent within each group. The MSAT, pre-test of algebraic ability, and number of years of high school mathematics all indicate that the experimental sections were more capable and more experienced mathematically than the control sections. However, the 37% in the "remedial" category with respect to university math history indicates that a larger percentage of students in the control group had taken and passed the non-credit remedial algebra course prior to enrolling in the current course. This finding is also consistent with lower MSAT and a smaller number for years of high school mathematics. Given that more than 50% of the students in each group were taking the course as their first university mathematics course, the variability with respect to mathematical ability in each group could be quite large. Hence, mean scores on the pre-test for each group at approximately 5.0 are also consistent.

The attitude scale, a 10-item Likert scale administered both prior to instruction and at the end of the semester, assessed the students' view about learning in general, the role of the teacher, and whether students believe they learn better alone or working with other students. Prior to instruction, both groups favored working with other students as the preferred way to learn mathematics. At the end of the semester, students' responses to the question, "I found working in cooperative learning groups to be (please elaborate)..." ranged along a continuum from "extremely enthusiastic" to "helpful but..." to "not at all useful." The results of this question are indicated in the following table:

Table 2: Responses concerning Cooperative-Group Learning

Categories of Responses	Extremely Helpful	Helpful, but...	No Comment	Not Useful
Control Group	53%	28%	5%	14%
Experimental Group	60%	13%	7%	20%

Following instruction, both groups still favored working in cooperative-learning groups. However, significant differences emerged with respect to the role of the teacher. The number of experimental group members who strongly agreed that "the role of the teacher is to facilitate learning" increased dramatically. Members of the control group remained ambivalent concerning the role of the teacher. The comments of many of the students with respect to working in cooperative-learning groups are represented by this excerpt from the Attitude and Cooperative-Learning Assessment.

> At first I didn't like working with other people because I usually study and work alone in order to memorize and teach myself information (which is hard to do with others). But by the end of the semester I enjoyed working with my group and I studied with 3 others for the final exam.

With respect to whether working in cooperative-learning groups effected problem-solving strategies, the student writes:

> I have listened to and heard other strategies and learned new solving and thinking patterns.

In order to investigate students' problem-solving strategies and their discourse with respect to a particular task, Task II was designed to investigate the connections between algebraic and graphical representations. This task or lab began with an Objective, Pre-Lab Activity, and Narrative Situation. The Objective and Pre-Lab Activity sections indicate that the purpose of the lab is to use the computer capabilities to explore the relationships between various representations and to form connections between specific representations: narrative, algebraic, tabular, and graphical. The Narrative Situation reads as follows:

> Tim brought a portable TV to school in September of his freshman year. He paid $280 and noted that the rate of depreciation for the TV was $5 per month. Anne brought a CD-player to school with her for which she paid $350. She noted that the rate of depreciation was $10 per month. There are several questions which we want to answer:

1. When will the TV and the CD-player be worth the same amount of money?
2. How much money will each machine be worth at this point in time?
3. When will each machine be worth $0?
4. If we assume that both Tim and Anne finish school in approximately 45 months, how much will each machine be worth when they graduate?

The lab continues by providing a Tabular Representation through partially completed tables of values for each machine. Students are asked to complete the tables for the given number of months (up through 36) and determine "the equation that describes the value, y, of each machine as a function of the independent variable x (time in months." Thus, they were asked to construct the Algebraic representation from the Tabular representation. Students were then asked to use the computer to graph each equation and adjust their viewing windows so they could easily examine both graphs. At this point in the lab, the students were asked to answer the sequence of questions posed above by using the TRACE feature of the graphing utility. After determining the answer to each question, they were asked, "what feature of the graph provides the solution to this question?" Now, after answering the questions graphically, the students were asked to use the algebraic representation to answer the questions. For each of the original questions they were asked:

1. What strategy can you use to solve this algebraically?
2. Why does this strategy work?
3. Now use the strategy to determine...

The final question of the lab was new and asked them to determine when each machine will be worth $100. They were instructed to use either the graphical or algebraic representation and to explain their solution clearly and concisely in complete sentences.

Analysis of both the videotapes of students working in cooperative-learning groups and their written work revealed four distinct connection strategies. These strategies, which are described below, seem to follow a hierarchy from a conceptual integrated understanding of the connections between the two representations to an algebraically correct strategy that seemed to indicate little understanding of the connection. The strategies are first listed and then described and illustrated.

1. Set Equations Equal (Conceptual): This strategy described those students with the most sophisticated understanding of the connections between the algebraic and graphical representations. These students indicated through their explanations that they understood that the algebraic solution of "setting the two equations equal to each other or a specific value" was the same as looking for a particular graphical feature, i.e., point of intersection (x-coordinate or y-coordinate).

Examples of Strategy:

Hold y constant and solve for x... holding y constant is the same as looking for the point on the graph.

Set the 2 equations equal because if you set the equations equal you can find where the 2 lines intersect on the graph and then are equal at that point.

Set the two equations equal to each other...it works because $y = y$ is also the intersection point of the lines $-5x + 280 = -10x + 350$.

2. Set Equations Equal (Procedural): This strategy described those students with a truly procedural understanding of the process. They knew that this would give them the correct answer and could correctly execute the procedure. However, they did not indicate in their explanations that they understood the connection between the algebraic and graphical solutions.

Examples of Strategy:

Set the equations equal to each other because when you set the equations equal to each other the solution will work out with the correct variable. Because we made both sides equal to each other.

3. Substitution of Numerical Value: This strategy described those students who understood from the graphical solution to check that the algebraic solution is correct. These students simply substituted the X obtained from the graph into each equation and noted that the result was the same or substituted for both the X and Y and noted that the result was a true statement.

Examples of Strategy:

Because x = 14 for both and you plug x = 14 into both original equations.

Plug in the #'s for both x and y from the chart...all we need to do is put them into both equations.

Substitute the numbers in for x...it works because you get the same answer.

4. Solve a System of Equations: This strategy described those students who understood that solving a system of equations by eliminating one of the variables provided the solution to the question concerning when the two machines would be worth the same amount of money. While this represents a relatively sophisticated algebraic understanding, there was no indication in their explanation that they understood this to be the point of intersection of the two lines.

Examples of Strategy:

Add equations together to get rid of the x-value and solve for y then plug back into original equation.

System of equations...it works because I learned it in class.

Eliminate one of the variables...find an answer for one and plug it into the equation.

The following table indicates the number and percentage of students that used each strategy.

Table 3: Connection Strategies

Strategies Group Responses	Set Equations Equal Conceptual	Set Equations Equal Procedural	Substitute Numerical Value	Solve a System of Equations
Control n = 30	6.5 or 22%	15 or 50%	7 or 24%	1.5 or 4%
Experimental n = 78	29 or 37%	14 or 18%	24 or 31%	11 or 14%
Total n = 108	35.5 or 33%	29 or 27%	31 or 29%	12.5 or 11%

In addition to examining the types of strategies that students used we also asked them whether working in cooperative learning groups has "**increased, decreased**, or **not changed**" the strategies I would use to solve problems. The following table indicates the results.

Table 4. Responses Concerning "Working in cooperative-learning groups has..."

Response Group	Increased	Not Changed	Decreased
Control Group n = 30 4 did not respond	14 or 54%	12 or 46%	0 or 0%
Experimental Group n = 78	33 or 42.5%	44 or 56.5%	1 or 1%
Total n = 108	47 or 45%	56 or 54%	1 or 1%

Analysis of Video and Audiotapes

With respect to the analysis of the video and audiotapes we found the following behaviors:

1. Students' attendance was much higher and they asked more questions in the CLG than in the control group.

2. Students motivate each other to do the assignment.

3. Individuals in CLG's seemed to exert pressure on one another to achieve or complete their activities. They asked each other questions to see whether each member understood what was going on.

4. During either the problem-solving activities or the computer labs, members of the CLG's assumed well-defined roles - recorder, typing information into the computer, asking questions, monitoring their progress - none of which were assigned by the instructor.

5. Students communicated to each other mathematically. They explained mathematical concepts and demonstrated (showed) different strategies to solve the problem.

6. When students started working on their activities, those in the CLG did not verbalize immediately. They briefly discussed an "overview" of the activity and paused to question each other. Those students in the "control" group started talking right away.

Discussion and Conclusions

In the laboratory activities, students were asked to estimate graphically, identify the graphical feature they utilized to answer the question, algebraically answer the question, and describe how the graphical and algebraic representations were related. Results of the written responses were significantly higher for the students in the cooperative-learning groups than those in the control group. Furthermore, those students in the cooperative-learning groups, like those in the "controversy group" identified by Smith, Johnson, and Johnson engaged in the type of mathematical discourse that would enable them to form connections between graphical and algebraic representations. In particular, their discourse indicated a willingness to persist with the problem and to explore alternative mathematical solutions.

Students in the cooperative-learning groups still exhibited difficulties explaining the connections between mathematical actions and/or processes and the mathematical concepts. Although, their work with the four tasks focused on forming and explaining these connections, the standardized assessments (multiple-choice questions with some free-response parts) did not emphasize forming these connections.

The control group did much better in the course than the experimental group, although the baseline assessments would indicate otherwise. One possible explanation is the differences between the two groups in terms of the number of students who took and passed the remedial algebra course prior to taking the current course. Another possibility centers around the differences between the problem-solving activities and/or labs and the standard exam questions. As with many college algebra courses at universities throughout the country, the end-of-course knowledge and skills that are emphasized still center on symbolic manipulation. As long as this is the case, those students who demonstrate problem-solving skills and mathematical understanding without symbolic manipulation skills will not realize their mathematical potential.

References

Artzt, A.F. & Newman, C.M. (1990). *How to use cooperative learning in the mathematics classroom*. Reston, VA: National Council of Teachers of Mathematics.

Brechting, M.C., & Hirsch, C.R. (1977). The effects of small-group discovery learning on student achievement and attitudes in calculus. *American Mathematics Association of Two-Year Colleges Journal, 2*, 77-82.

Chang, P.T. (1977). On relationships among academic performance, sex difference, attitude and persistence of small groups in developmental college level mathematics courses (Doctoral dissertation, Georgia State University). *Dissertation Abstracts International, 38*(7), 4002A.

Dees, R.L. (1991). The role of cooperative learning in increasing problem-solving ability in a college remedial course. *Journal for Research in Mathematics Education, 22*(5), 409-421.

Davidson, N. (1971). The small-group discovery method of mathematics instruction as applied in calculus (Doctoral dissertation, University of Wisconsin, 1970). *Dissertation Abstracts International, 31*(11), 5927A (Tech Rep. No. 168). Madison: Wisconsin Research and Development Center for Cognitive Learning.

Olsen, J.C. (1973). A comparison of two methods of teaching a remedial mathematics course at the community college (Doctoral dissertation, Utah State University). *Dissertation Abstracts International, 34*(120), 7522A.

Schoenfeld, A.H. (1987). What's all the fuss about metacognition? In A. Schoenfeld (Ed), *Cognitive Science and Mathematics Education.* Hillsdale, NJ: Lawrence Erlbaum.

Shaughnessy, J.M. (1977). Misconceptions of probability: An experiment with a small-group activity-based, model building approach to introductory probability at the college level. *Educational Studies in Mathematics, 8*(3), 296-316.

Slavin, R.E. (1995). *Cooperative Learning.* Boston, MA: Allyn & Bacon.

Smith, K., Johnson, D.W., & Johnson, R.T. (1981). Can conflict be constructive? Controversy versus concurrence seeking in learning groups. *Journal of Educational Psychology, 73*(5), 651-663.

Treadway, R.T. (1983). An investigation of the real-problem-solving curriculum in the college general education mathematics course (Doctoral dissertation, University of North Carolina at Greensboro). *Dissertation Abstracts International. 45*(1), 108A.

SECTION 2

EDUCATIONAL TECHNOLOGY

4

COLLEGE ALGEBRA:
TRADITIONAL INSTRUCTION
VERSUS INSTRUCTION VIA VIDEO TAPE

Richard Mitchell

Background

Most universities offer a college algebra course that essentially duplicates the content from 9th grade algebra. During the fall semester of 1995, the University of Wisconsin-Stevens Point (UW-SP) offered 7 sections of college algebra; three years ago we offered 12 sections. The reason for this decrease is simple: the entrance requirements to the university have been strengthened. In fact, starting in the fall of 1996 entering freshmen will be required to complete three years of high school math. I anticipate that enrollment in college algebra will continue to decrease and that UW-SP will move away from the business of teaching introductory algebra. (Note: There will likely always be a need for college algebra since non-traditional students are not required to satisfy the same entrance requirements. It is also interesting to note that as the entrance requirements are strengthened, the percentage of non-traditional students enrolled in college algebra will increase.)

In addition, the administration on the UW-SP campus is interested in determining the extent to which technology can be used to free-up faculty time in the classroom. If instruction via technology is at least as good as traditional instruction, then perhaps technology can be used to re-place/supplement instruction. This extra time could be used for endeavors such as curriculum development, professional development, and scholarly activity or re-channeled back into the classroom through activities such as extra tutoring, collaborative learning, and enrichment activities.

The question arises as to the best way to teach college algebra to a decreasing population of students in need of such a course while at the same time incorporating strategies to increase the number of students we can serve, without a commensurate increase in faculty time and effort.

Multimedia Technology in the Algebra Classroom

Although multimedia computer technology is relatively new, it is fast becoming a standard in the market place. Multimedia computer technology may include text and exercises that you would normally find in a printed book, in conjunction with computer algebra systems (e.g., Derive, Maple, and Mathematica) that allow students to visualize and explore algebraic relationships. Furthermore, multimedia technology may include on-line testing and grading. Finally, with the advent of MPEG technology, full-screen video can now be incorporated (e.g., a full-length movie will fit onto a single CD).

Since multimedia technology combines the traditional textbook, symbolic manipulation software, and VHS quality audio/video into a single package, portions of the college algebra sequence might be better taught via the computer. It is important to note that the technology does not replace the instructor, but rather allows for the modification of the instructor's role. In the modern, technology-rich learning environment, the instructor becomes the facilitator of learning rather than the disseminator of information. The most important role is that of providing personal assistance to students: office hours, one-on-one assistance, small and large group discussions. In the future, much of this might even be accomplished on-line, via the World Wide Web, which plays well for "distance learning."

An extensive search was undertaken to determine the availability of materials appropriate for algebra instruction. Although I was primarily interested in multimedia materials, the search was not limited as such. Since the focus of this project was faculty productivity, only products that could be completed by students with minimal faculty assistance were sought. For example, graphing calculators were considered, but then ruled out because these materials require the amount of faculty assistance to increase

I found few companies that were developing and/or using multimedia materials for mathematical instruction, but I do expect the number to grow in the near future. I was not able to locate any multimedia materials for stand-alone algebra instruction. Although there is a great deal of computer materials available for enrichment, very little is classed as stand-alone. Innovative, stand-alone, multimedia algebra materials should be available sometime toward the middle of 1996; the most promising company is Addison Wesley Interactive.

Video Tape Instruction

In the course of investigating the availability of multimedia materials for algebra instruction, I learned that many textbook companies provide video tapes that are tied to their sequence of materials. I viewed the video tapes from several textbook companies and found them to be excellent. There are several advantages to video tapes over traditional classroom instruction:

- Instruction is not tied to a particular time; students can view the video tapes at various times of the day or week.
- Portions of the tape can be rewound and replayed.
- Students can work ahead - useful when they know that they will miss a class.
- There is consistency in the instruction — all students receive the same instruction.

A distinct disadvantage of video tape instruction is the lack of interaction between the students and the instructor. For instance, that level of concern and attentiveness that students feel because the instructor may call on them to answer a question.

Since full-screen video will be a key component of future multimedia developments, I decided to determine the effectiveness of using video instruction in the classroom. Working closely with my department chair, I began by developing a preliminary study with the treatment (i.e., video tapes) scheduled for the full length of the course. Unfortunately, problems surfaced with the non-tenure-track instructional staff who typically teach the introductory courses. When they heard "video instruction" their first thoughts were "replace instructors." It was a great challenge to overcome their concerns of being replaced by video instruction. Eventually they realized that the project was not about using technology to replace instructional staff, but rather, the idea was to redefine the role of the instructor in a modern, technology-rich classroom. We felt that it was important for the instructional staff to be active in this redefinition.

Video Tape Experiment

Eventually, two instructors agreed to participate in the project. Early concerns of the two instructors included the following:

• If instruction is primarily by way of video, what will the instructor do? Except for lecturing, the instructors were not aware of other ways to use class time to facilitate student learning.
• They were not comfortable with the idea of using videos in a mathematics classroom.

In the end, we decided to taper back the experiment and utilize the videos in only a portion of the course. Important components of the video project included the following:

• We utilized three sections of Math 100, College Algebra, during the fall semester of 1995. These were two-hour courses that met for eight weeks.
• We used the video tapes supplied as ancillary materials with our current textbook (*Intermediate Algebra*, Prentice Hall).
• The videos were used in two units (i.e., two weeks), rather than the full eight weeks.
• Students watched the videos outside-of-class. The video tapes were available in the library and also for short-term check out.
• The course met four times a week. During the weeks that videos were used, the whole-class met for two periods. Rather than lecture, class time was primarily used for small group processes (e.g., guided practice

and collaborative learning). The instructors were available during the other two periods (either in their offices or in the classroom).

- Students were provided with a sheet that included the objectives (i.e., what should they learn) and the assignment (i.e., reading, videos to watch, homework). This was important because some students did not come to class during this time.
- Students self-reported the time that they spent watching the videos and kept a journal of their thoughts, observations, and suggestions (see Appendix A).
- One instructor offered extra credit for watching the tapes; the other instructor did not.
- The instructors kept track of the time that they spent helping students outside-of-class. We wanted to know if this time increases because primary instruction is via video tapes.
- Student evaluations were not used for personnel matters.

What We Learned

- Although both instructors were reserved in the beginning, in the end, they were strong advocates of using the videotapes in the classroom.
- The two instructors were especially enthusiastic about using collaborative learning in class. Because of time restrictions, in-class, small group learning is a strategy that the instructors had never used in a college classroom.
- There was not a significant difference between video learning and traditional learning as measured by the common final exam. We compared the final exam scores from the fall 1995 semester to the final exam scores given in the three previous years. This result is not surprising since the treatment period was so short.
- Extra credit went a long way in motivating students to view the video tapes. When extra credit was not offered, students participated minimally.
- Office hour appointments did not increase significantly during the video portions of the study.
- The two instructors reported that the videos were especially helpful for
  a) non-traditional students;
  b) students with math anxiety;
  c) students with physical handicaps;

d) students who miss class;
e) students who have difficulty writing and listening at the same time; with the videos they can primarily write while in class and use the videos to listen to (or vice versa).

- The students were satisfied with the quality of the video learning, felt that the video learning was a good complement to the instructor's teaching, and found it easier to understand the concepts presented in class after watching the videos (see the charts in the Appendix B).

- It would have been helpful to have had a video study guide for the students. For example, a list of topics/ideas to look for when watching the videos.

- It might be useful to have the videos available at text rental; in the future this might be CDs.

- If the redefinition of the instructor's role in the classroom is to include collaborative/cooperative learning, the instructional staff will benefit greatly from professional development activities in this area. Although significant changes, both in content and methodology, are occurring in the high school algebra classroom, the college algebra classroom appears much as it did in the early fifties. If technology is used to free up time in the classroom, the instructional staff need to be aware of research-based, instructional strategies that can be used to facilitate student learning.

- Future textbook adoptions should give special attention to the availability of ancillary materials such as video tapes.

- The term "video learning" holds strong negative connotations by some students (especially non-traditional students) and some instructional staff. This is likely the result of activities they participated in, or are aware of, from the 70s when there was an explosion of technology-based learning (e.g., computer assisted instruction). Great care needs to be given to the process of introducing a new technology or instructional practices into the classroom.

Conclusions

- A major goal of this project was to determine how video learning will be received on this campus. We found that the video learning that took place in this project was well received by both the students and the instructors involved. Because full-screen video will be a key component of future multimedia developments, this finding comes as good news.

- We determined that technology could be used to free-up faculty-time in the classroom. For this project, we re-channeled that extra time back into the classroom through collaborative learning activities for the students. In the course of doing this, we began the process of rethinking the instructor's role in a modern technology-rich classroom.

Appendix A
Student Responses to the Videos

1. Watching the video tapes proved to be extremely helpful. When I first began watching the tapes, I would do so right after my math lecture. This seemed to work adequately; however, I sometimes found myself lost during lecture. What I found to be most helpful was to watch the tape prior to going to class. This resulted in the everyday lectures being much easier to follow and understand. The class material not only makes more sense but also resulted in more confidence in my abilities. Other classmates that had not viewed the tapes seemed to struggle to follow and understand the same material, which I was following fairly easily.

2. They helped me a lot. The material is well covered on the tapes but I wouldn't replace the teacher for them. The tapes worked best for me when I already had a grasp on what I was doing. They were great reviews. I would recommend people to watch them. They were pretty short, but give a lot of information.

3. The thing that I didn't like was that you couldn't ask questions, It was very impersonal.

4. It helps to go through the material in class first, then review with the video.

5. I started to do my homework before I watched the video. I was having trouble so I decided to stop and watch the video and it was easier for me to understand.

6. Watching the videos really helped with the assignments. They were a review, but it helped because it seemed like I forgot how to do it.

7. I did not really get the ball rolling until I watched the video. It helped because I didn't quite catch it all in class. The videos explained it really well.

8. If I were to do this again, I would actually watch the videos before learning it in class.

9. I watched them 2x — They *really* helped me.

10. The videos present it a little differently — get another point view on how to solve problems.

11. Overall, the videos were helpful. But I find it more difficult to learn from a lady on television. Also, it made me feel a little dumb watching people walk by starring at my videos screen seeing a lady teaching me basic algebra. But I guess the biggest thing that bugged me about the videos is that they lacked the interaction that you can get in a classroom.

12. First time I watched the video, it was very confusing. But I watched it over again and it makes more sense.

13. This section was very hard for me to understand because it was brand new to me. But its a lot better now because I did the homework, watched the video, and had the class discussion.

14. This stuff is very confusing to me. I think that it was good that I watched the video before going to class.

15. I think that the videos are very beneficial. They certainly helped me a lot. Some videos offered a good review, and others helped clarify things (by going step-by-step).

16. I think that seeing the video will prepare me for when we go over it in class. I actually understood what she was doing on the video. It made sense!

17. I like to view the video before we discuss it in class. It makes comprehension (during class) of the skills easier.

18. I found the video extremely helpful in several ways. I was able to follow the formula definition in the text as the video instructor explained them. Also, when the video instructor choose examples from the text, I could stop the video and work the problem at my own pace and then restart the video to go step-by-step through my work.

19. I think that the video was helpful. It is nice to hear another teacher's way of teaching the material.

20. I think it is a pretty good video, but I personally like having a teacher that I can ask questions to. I had video school from 5-8th grade and I don't really want to go back to it.

21. The videos are from my perspective (nontraditional student), extremely helpful. The use of the tape allows the emphasis of learning to be concentrated on concepts that are difficult to grasp.

22. The videos are not as exciting or humorous as my professor, however, very effective in a positive way.

23. I feel the videos are helpful if you have the time to view them. The video instructor is very thorough and take things at a good pace. It's just hard to find time to get in to watch them.

24. The video is good, but will never replace the teacher! It is a great review, re-explanation, and offers a chance to work ahead.

25. It is good to have the videos if you want to work ahead, or don't understand the material in the book and need it explained again.

26. The videos were easy to follow and understand, but I didn't always have the time to go to the library to watch them.

27. The video was very helpful for this course because we could watch it before class. If we have any problems then we could ask the teacher during class and don't waste time on things we already know.

28. I'd rather by taught be a professor, then a machine.

29. The videos are very helpful. But it would be better if the library was open longer hours or if we could take the videos home and view.

30. I think that I learn more in class then watching the videos. It's easier in class.

31. The videos are O.K. if you miss class, but I don't think you should assign them. It's easier to come to class for an hour then to go to the library and watch a video.

32. The video offer a little different style of teaching sometimes. It's nice to be able to control the rate of learning by fast forwarding through familiar areas and rewinding when unsure about something. The flexibility of viewing times is also nice.

Appendix B

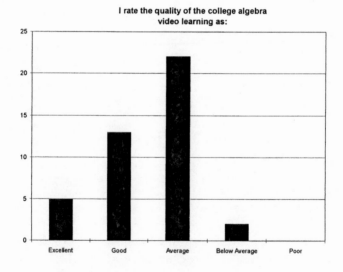

I rate the quality of the college algebra video learning as:

Video learning is a good complement to my professor's teaching.

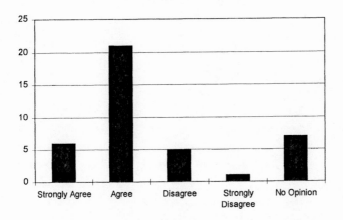

After watching the videos, I find it easier to understand the concepts presented by my professor in class.

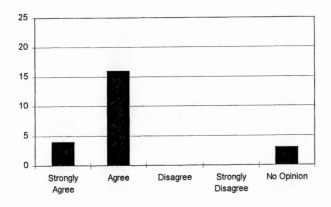

5

THE SPREADSHEET AS A LEARNING TOOL
IN CALCULUS

Louise M. Berard

An electronic spreadsheet provides a convenient means for conveying the intuitively dynamic aspect of various "limiting processes" studied in first-year calculus. As a first example, let us consider how in beginning calculus we introduce the concept of limit of a function at a point — e.g.,

$$\lim_{x \to 2} 3x - 1.$$

In first-year calculus, we intuitively regard such a limit as involving a dynamic aspect; we use phrases such as "x approaches 2." Initially, we ask students to simply evaluate $f(x) = 3x - 1$ at values of x chosen successively closer to 2. Using a spreadsheet eliminates the tedium of performing the calculations, so that students can focus on the behavior of the function values as x gets closer to 2.

For an in-class demonstration, the instructor may prepare in advance a spreadsheet with selected x-values already entered in one column. Then during class the instructor may enter the formula for f(x) in the top cell of a second column, and paste this formula into the remaining cells one at a time, proceeding towards 2 first from above then from below. The relative address in the formula is automatically updated to refer to each x-value in turn. This is more convincing than merely examining a table whose entries have all been entered in advance, and requires less class time than using a hand-held calculator to complete a table on the blackboard. One could use a similar method when introducing the derivative as a limit of quotients.

	A	B	
1	x	f(x)=3x-1	
2	1.9	4.7	← entered as formula 3*A2 - 1
3	1.95	4.85	(then copied into the
4	1.98	4.94	remaining cells of column B)
5	1.99	4.97	
6	1.995	4.985	
7	1.999	4.997	
8	1.9999	4.9997	
9	2.0001	5.0003	
10	2.001	5.003	
11	2.005	5.015	
12	2.01	5.03	
13	2.02	5.06	
14	2.05	5.15	
15	2.1	5.3	

To view graphically the behavior of these tabulated function values, it is convenient to use the XY graph type provided by some spreadsheets. The advantage of the XY graph type is that the x-values can be used to determine a uniform scale for the horizontal axis. (Most graph types in a spreadsheet treat cell values merely as labels to be equally spaced along the axis.) In the following graph, only the discrete data points are plotted.

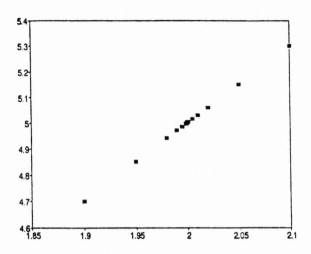

When we discuss the formal definition of limit, again we communicate a dynamic aspect, namely, a dynamic interplay between ε and δ. We imagine a value for ε being specified, then finding a corresponding value for δ; another value for ε specified, another value for δ found; and so on. A spreadsheet's graph annotator facility allows us to represent graphically the relationship between ε and δ. Consider the preceding example, the limit of f(x) = 3x − 1 as x approaches 2. In the spreadsheet itself:

- Use one cell to hold a value for ε; by hand, determine how δ depends on ε; in another cell, enter the formula to compute δ (e.g., our example is a linear function with slope 3, so if e is entered in cell A1, then the cell for δ would contain the formula A1/3);

- In one column, enter the x-values 1, 2 − δ, 2, 2+ δ, 3;

- In another column, enter the formula to compute corresponding values of f(x).

Plot x versus f(x) on an XY graph; using the graph annotator, link horizontal and vertical lines to the points corresponding to 2 − δ and 2+ δ (once the lines have been linked to data points, they will change appropriately as ε is changed). The picture for ε = 0.5 appears as follows.

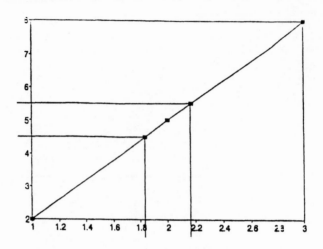

For our next example, we consider Riemann sums for a function on an interval — e.g.,

$$f(x)=x^2 \text{ on } [0, 1].$$

Let us consider a regular partition with n subintervals, and with f evaluated at the right endpoint of each subinterval. The nth Riemann sum is

$$(*) \quad \frac{1}{n}\sum_{k=1}^{n}\left(\frac{k}{n}\right)^2 = \frac{1}{n^3}\sum_{k=1}^{n}k^2$$

A sequence of Riemann sums for the given function can be computed with a spreadsheet as follows:

• Enter successive values of n in one column (in the spreadsheet below, cell A_n holds n);

• Use another column to hold the corresponding squared integers (cell B_n holds n^2);

- In a third column, enter the second formula in (*) for the Riemann sum (cell C_n holds the sum of entries B_1 through B_n, divided by the cube of entry A_n).

absolute address

A	B	C		
1	1	1	1.0	entered as @SUM(B1..B1)/(A1^3),
2	2*	4	0.625	then copied into the remaining
3	3	9	0.518519	cells of column C
4	4	16	0.46875	
5	5	25	0.44	
6	6	36	0.421296	
7	7	49	0.408163	
8	8	64	0.398438	
9	9	81	0.390947	copied as @SUM(B1..B9)/(A9^3)
10	10	100	0.385	

.
.

100   100 10000   0.33835

*   Entry A2 is entered as formula 1+A1,then copied into the remaining cells of column A.

Observe that, as n increases, the nth Riemann sum provides a better approximation to 1/3, which is the value of the definite integral

$$\int_0^1 x^2 dx.$$

Alternatively, if we wish to emphasize the area interpretation of the definite integral, we could tabulate individual terms for a single Riemann sum, and construct a 2-dimensional bar graph to represent individual terms by rectangles. Consider the preceding example.

- For a fixed value of n, enter the values 1/n through n/n in a column (say, cell A_k holds k/n) — e.g., for n = 10, use the "fill block" command with start value 0.1, increment 0.1, and stop value 1;

- Use another column to hold corresponding function values (cell B_k holds $(k/n)^2$);

- In a separate cell, enter the first formula in (*) for the Riemann sum (cell C_1 holds the sum of entries B_1 through B_n, divided by n).

Draw a 2-D bar graph, specifying column A as the series for the horizontal axis and column B as the series for the vertical axis; use the same scale on both axes, and adjust the bar width to be as large as possible (the graph annotator can be used to sketch the curve):

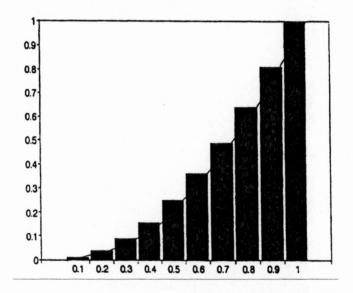

As our last example, we consider partial sums of an infinite series — e.g.,

$$\sum_{k=1}^{\infty} 2^{-k}.$$

It is straightforward to use the spreadsheet to compute partial sums as follows:

• Enter successive values of the index in one column;

• Use another column to hold the corresponding terms of the series;

• In a third column, enter the partial sums (in the spreadsheet below, cell C_n holds the formula $C_{n-1} + B_n$).

	A	B	C	
1	1	0.5	0.5	← entered as +B1
2	2	0.25	0.75	← entered as +C1+B2
3	3	0.125	0.875	(then copied into the re-
4	4	0.0625	0.9375	maining cells of column C)
5	5	0.03125	0.96875	
6	6	0.015625	0.984375	
7	7	0.007813	0.9921875	
8	8	0.003906	0.99609375	
9	9	0.001953	0.998046875	
10	10	0.000977	0.9990234375	
.				
.				
49	49	1.78E-15	0.999999999999998	
50	50	8.88E-16	0.999999999999999	
51	51	4.44E-16	1.000000000000000	

By studying the partial sums as n increases, students gain insight into the meaning of "sum" of an infinite series. Furthermore, they note the distinction between the sequence of terms and the sequence of sums Students could create a spreadsheet as above for several different divergent series and compare the rates at which the partial sums grow.

These examples illustrate ways in which an electronic spreadsheet can be used to enhance the learning of calculus. Spreadsheets are easy to use, and many students have prior experience with spreadsheets from other courses. A spreadsheet is particularly appropriate for calculus courses designed for business students, because such students are afforded the opportunity to employ a familiar tool in unexpected ways.

Student Reaction to Spreadsheet Use

Based on informal observations that I have made during the past few years when I have used spreadsheets to illustrate concepts in calculus, I conclude the following: Students are becoming more familiar with spreadsheets as the use of spreadsheets increases throughout the undergraduate curriculum. When I first started exploring mathematical uses of spreadsheets, many students had not worked with them in other courses, so I had to spend significant time in my classes discussing basic features of a spreadsheet. This is no longer the case. Indeed, I see this general familiarity with spreadsheets, as well as their wide availability, as definite advantages for students.

Regarding the specific examples presented in this article, student reaction seems more positive for the later examples on series summation than for the earlier examples on limits of functions. For instance, when using a spreadsheet to generate specific partial sums of an infinite series, students seem to grasp the notion of "sum as limit" more readily than did students in my pre-technology classes. On the other hand, in presenting my first example on the intuitive notion of limit, sometimes I use a spreadsheet and other times the table feature on a graphing calculator, with little apparent difference in student response.

6

COMPUTER BASED DIAGNOSTIC TESTING AND SUPPORT IN MATHEMATICS

Stephen Hibberd
Andrew Looms
Douglas Quinney

1. Introduction

For students entering Higher Education (HE) courses in Science and Engineering there is always some level of pre-requisite assumption and reliance on prior knowledge in a range of topic areas and mathematical skills. Such courses also tend to recruit large numbers of students with a rich diversity of intake qualifications and prior experiences. The need to assess on entry the current active ability of students to any course is crucial. To be able to do so rapidly and effectively, and provide suitable student centred support, remains an ongoing challenge. The experiences of the authors in the implementation of computer-based Multiple Choice Questions (MCQ)-based tests and courseware materials in mathematics confirms that existing computer-based materials and technology is available to meet the diverse needs of students but is not yet universally recognised nor exploited.

In this article we argue for the need to establish an educational strategy that begins with diagnostic testing, provides suitable support material which is a symbiotic relationship between computer based material and tutorial support, and assessment.

2. Diversity of Student Prior Experience

Besides its importance as a subject specialisation in its own right, mathematics is prominent as a core skill required by students of engineering and science and increasingly of management studies and social sciences. However, there is currently ongoing concern at the level of mathematical capabilities of students entering Higher Education (HE) and the level of preparedness to proceed successfully into various undergraduate programmes (LMS Report (1995)). Mathematics is a subject that, at the outset, relies and builds on existing knowledge and techniques. Deficiencies in any significant prerequisite material can be crucial for subsequent understanding and progression. Recently there have been substantial changes within all Examination Boards within the UK towards modularity. Furthermore, there has been a substantial decline in student numbers arriving at University with 'traditional A-level mathematics' qualifications, taken typically at age 17-18. In particular there is a noted decrease in students taking two Advanced level subjects in Mathematics ('Double Maths'). (Kitchen (1996).) This has resulted in University staff dealing with a constantly evolving student knowledge and skills base. In addition, when a student arrives at university, it is difficult to know exactly which topics have been covered and in what detail. For example, Table 2.1 shows the coverage of non-core topics at A level in the various examination boards. Furthermore, the year 2000 has been seen even more changes in the structure, curriculum and assessment of students when further changes to A-level are implemented. The implicit assumption that all students start on a level playing field in terms of background knowledge is unlikely to be true even at the most prestigious universities.

Table 2.1: The Variation in Non-Core Topics at 9 A-Level Boards in the UK

	AEB	CAM	CAM Mod	ULE	NEAB P2	NEAB P4	NEAB SMP	OXF	OXF Mod	O&C P2	O&C Mod	NICC Mod	WJEC	SEB
Num. soln. equations	*	*	*	*	*		*	*	*		*	*		
Polynomial fns.	*	*		*	*	*			*	*	*			*
Rational functions	*	*		*	*	*			*	*	*			
More series								*		*				
Perms and Combs.		*	*	*			*	*	*		*	*	*	
Binomial series	*	*		*			*		*		*	*	*	
Exp/Log series	*	*	*				*	*	*	*				
Straight lines	*	*	*	*			*	*						*
Circles/Spheres	*	*	*				*	*						*
Experimental laws	*	*	*	*	*									*
Sketch curves					*	*	*							*
More trigonometry	*	*	*							*	*	*	*	*
Vectors		*			*		*	*	*	*	*	*		*
Vectors & Geometry	*	*			*		*	*	*	*	*	*		*
Matrices									*			*	*	
Complex numbers								*		*		*	*	
Complex graphs								*		*		*	*	
More differentiation	*	*	*	*	*	*	*		*	*			*	*
App differentiation	*	*	*	*	*	*		*	*		*	*	*	*
More integration		*	*		*	*						*	*	
App integration		*	*	*	*	*	*	*	*			*		*
Differential eqns.	*	*	*	*	*			*		*	*	*		*
Num. quadrature		*	*				*	*				*		

There is a perception in the UK that a decrease in the mathematical ability of students entering university is due to falling numbers of students taking A-level mathematics. There are, however, many other factors that influence the skills possessed by students entering HE, and indeed the number of such students. Table 2.2 shows the number of students who opted for a single A-level in mathematics in recent years and those which obtain a pass. Although the path that students take may have changed, the numbers are remarkably consistent.

Table 2.2: Students Entered/Passed at A-level ('000)
(Source DfEE)

	School Entries	School Passes	FE Entries	FE Passes	Total Entries	Total Passes
1991/92	41.8	32.2	12.0	7.0	53.8	39.2
1992/93	33.5	27.2	18.9	13.1	52.4	40.3
1993/94	32.2	27.1	17.8	13.0	50.0	40.1
1994/95	32.6	28.8	16.8	13.1	49.4	41.9

The table seems to indicate that there does not appear to be a decline in the numbers taking A level; but that their distribution is very different. Table 2.3 gives a revised table that distinguishes between those students who take one or two A Levels. It was the latter group who once made up the major entry into single honours mathematics courses.

Table 2.3: A-Level Mathematics Candidates
(1980 & 1985 England, 1990 & 1995 England & Wales)
Kitchen (1996)

	Single A Level	Double A Level
1980	54.1	13.4
1985	63.9	11.9
1990	62.6	6.9
1995	51.0	5.5

The most notable feature is the significant decline in double A Level candidates even without considering the actual number of students who achieve a passing grade. Finally, Table 2.4 gives interesting figures regarding the ever-present argument of changing standards.

Table 2.4: Students Obtaining Grades A-E at a Level in Mathematics (Times Educational Supplement, 1990-1999)

	A	B	C	D	E	Fail
1990	17.2	14.5	14.1	14.3	13.4	26.5
1992	20.1	14.6	14.2	14.0	13.0	24.1
1994	25.3	18.1	16.5	14.3	11.0	14.8
1996	26.7	19.3	17.0	14.1	10.9	12.0
1998	28.0	19.1	16.7	13.9	10.6	11.5
1999	28.3	19.0	16.5	13.8	10.7	11.1

However, these figures must be read with care as the entrants in later years are self-selecting and may be affected by other factors, further demonstrating the complexities of gauging intake abilities of students on qualifications alone.

The previous tables tell us that the overall capability of students entering higher education is now very different from even 10 years ago. This does not mean that the students are any less able; merely that their education up until entry is very different. With the present diversity of entry backgrounds, tutors and lecturers are finding it increasingly difficult to have confidence in either the range or depth of knowledge that can be assumed at the starting point of their courses. A poor remedy is to target the course to the lowest common knowledge base leading to frustration and lack of attainment of better prepared students. There is a clear need to establish a means of identifying individual student deficiencies and simultaneously promoting an environment in which, with tutor or other guidance, such deficiencies can be remedied on a person-to-person basis. Providing individual supplementary assistance through tutorials, workshops or CAL courseware can complement plenary teaching. The increasing availability of student centred computer based support material and diagnostic testing together provides a golden opportunity to create an integrated environment which holds the prospect of addressing the underlying problem within existing staffing resources.

3. Computer Based Diagnostic Testing

The need to rapidly assess the current ability of students on entry to any course is two-fold:

(i) Provide teaching and tutorial staff with a global assessment of the current active ability of each student on a chosen range of topics. Quantitative information attained is valuable in the following ways:
- provides a check on the intake ability of the student;
- indicates topics and levels of supplementary assistance that may be required;
- enables tutorial or clinic staff to target additional support at an early stage;
- helps in reviews of module curriculum and content reviews.

(ii) Provide students with useful individual feedback **before** problems escalate. Benefits to students include:
- early identification of possible black-spots or weaknesses;
- the pre-selection and identification of critical topics;
- guided support for remedial action;
- an assessment of their suitability for taking the chosen module.

Multiple Choice Questions (MCQs) are attractive to those looking for a faster way of assessing students arising from their ease of marking. Their simplicity is such that implementation for marking by computer, either through the use of Optical Mark Reading (OMR), or directly 'on screen' in real time is straightforward and provide for ready analysis and comparisons between groups.

3.1 Use of MCQ-Based Tests

A focus on diagnostic testing in mathematics was initiated through a mini-conference organised by Computers in Teaching Initiative (CTI) Mathematics organised and Government funded in the UK. (CTI Report (1996)). Currently most of the existing diagnostic tests are based on MCQs as an effective method with the specific advantages of simplicity of operation and implementation. Many diagnostic tests remain internally written, paper-based is common, as is the use of OMR, but increasingly the implementation of computer-based test based on Teaching and Learning Technology Project (TLTP) sponsored materials or proprietary test packages is becoming more widespread. The almost universal adoption of MCQs reflects their inherent strengths in providing effective formative feedback through the use of simple questions aimed at soliciting threshold knowledge. Whilst acknowledging their potential when used by a skilled practitioner to provide deeper insight, within a diagnostic framework the main demand rests on sampling rapidly a student's knowledge of a field of study. Further, the often-associated features of assessment security are not an issue and once constructed and validated they can be usefully reused continually to provide a year-on-year pattern of changes in student pre-requisite knowledge.

Within the mathematics fraternity it is apparent that if every institution were to develop independently its own tests then this may be both wasteful and inefficient. To this end the Heads of Departments of Mathematical Sciences in the UK (HoDoMS) has funded a WWW site giving information, contacts and case studies of existing diagnostic tests. **http://www. keele.ac.uk/depts/ma/diagnostic/**. It is, however, informative to review a number of testing mechanisms that are currently used in a variety of different formats.

3.2 Examples of Diagnostic Tests

The following approaches correspond to some areas of best practice and illustrate the keen activity in the areas of diagnostic testing.

CALM, Heriot-Watt University, UK

This diagnostic test was originally designed on paper by Professor John Hunter of Glasgow University in the late 1970s and further developed using Authorware, by staff at Heriot-Watt University. The diagnostic test is taken by all students taking mathematics and consists of 25 multiple-choice questions based at the level of Scottish Higher Mathematics and was designed to take 45 minutes. At the end of the test session the students are given an analysis of their performance which is also available to the course teacher. Further details are given by Beevers et al. (1992, 1996) or at the WWW address:- http://www.marble.ac.uk/maths/public/assessment.htm.

BP Mathematics Centre, Coventry University, UK

This is a diagnostic test that is processed by OMR. It is given to all students entering Engineering subjects to assess the mathematical skills and to help evaluate their individual needs. As the scores are recorded it is possible to analyse both individual students and long term trends. Lawson (1997) has published the results.

Mathlectics, Brunel University, UK

Mathelectics was developed by Greenhow using Question Mark Designer at the Department of Mathematics, Brunel University. An analogy is drawn between simple mathematical skills and single track events, grouped skills (such as problem solving) and the pentathlon, general skill and the decathlon and so on. It is aimed at non-principle mathematicians but its great advantage is its flexibility in designing more general questions that include problems that require hot spots but it is less flexible in its random capabilities. It comes with software to collect and analyse individual and group data. Further information is available at http://www. brunel.ac.uk/~mastmmg/.

Shell Test, University of Nottingham, UK

This test was developed as part of a CAL to investigate the knowledge for students entering a first year module in engineering mathematics. (Brydges S & Hibberd S, 1994, Hibberd S, 1996.) Of primary importance was simplicity of operation for students but it is also flexible enough for its 'shell' to be readily used for other diagnostic tests, self-assessment tests or as a possible grading mechanism. Keele University selected this test and further details are given below.

Diagnosys, University of Newcastle upon Tyne, UK

DIAGNOSYS is the product of TLTP in collaboration with four universities in the North East of the UK. The main features of the basic mathematics test implemented on DIAGNOSYS is that the questions asked are dynamic and based on over 90 'skills' organised into a network. The initial questions are based on prior qualification and following questions are selected using inference rules ("an expert system") so that only about

50 questions need to be answered. It takes about an hour and produces both student and tutor feedback. Currently, DIAGNOSYS investigates material at a relatively low level of mathematical sophistication but the shell which has been developed could be extended to more advanced material given suitable resources.

4. Computer Based Support

Gains made from the implementation of diagnostic testing are limited without also providing suitable "support material." Generally, a provision is sought which can be used provide an effective way of ensuring all students have individual access to the required core knowledge and appropriate support. This can be accomplished through tutors, clinics, supplementary lectures etc. but an increasing viable and cost-effective mechanism involving minimal staff resources is to utilise existing computer-based resources. Further, experience has shown that even though the weaknesses of individual students can be detected using diagnostic testing the restrictions of the timetable make it difficult to allot specific times when students can be supervised to ensure that any remedial work is carried out. However, since the test is computer based it seems appropriate to use the same medium to provide mathematics courseware support. Further, given that a diagnostic profile set against pre-determined objectives it is possible to provide automated links to CAL materials. In the UK the leading CAL packages addressing mathematics at the interface with university studies are *Mathwise*[1], CALMAT[2], CALM[3], Metric[4] and TransMath[5]. As the authors are members of the consortium that developed *Mathwise,* and which is also the most extensive of the packages, an outline is provided of the educational functionality together with an indication of level and width of support material that is available.

---

1. Mathwise — 50 Computer based modules developed by the United Kingdom Courseware Consortium (UKMCC).
2. CALMAT — Computer Assisted Learning in Mathematics based at Glasgow University
3. CALM — Computer Assisted Learning in Mathematics, Heriot Watt University.
4. Metric — TLTP project based at Imperial College, London.
5. Transmath — TLTP project based at Leeds University.

4.1 *Mathwise* Learning Environment

Mathwise is mathematics-based courseware based on some 50 modules of self-paced learning modules produced by a major group of academic authors from over 30 UK Universities funded by a government initiative (Teaching and Learning Technology Programme). The term courseware is used here to mean a form of learning related computer based material that is to some extent self-contained. A unique feature of *Mathwise* is that it aims to produce an integrated learning environment that is more powerful than simply a set of computer based learning materials and these features are explained below. *Mathwise* is readily available to UK Higher Education by FTP and on CD; Harding and Quinney give further details, (1996, 1997). This material is also being further refined and additional assessment added and made more widely available commercially (email: mathwise@nag.ac.uk).

A perceived advantage of using computer-based courseware is to exploit the attributes of

- Interactivity
- Student-driven
- Enrichment
- Exploratation.

The core element of the courseware is a module. The format of a module is fully specified in terms of screen layout and availability of navigation aids. Every module consists of a group of learning units that is equivalent to a mini-lecture with its own internal and external resources. There is no rigid formula and a learning unit may contain text, diagrams, animations, etc. Additional resources may either be smaller units, generally of single topic contextual material gathered into leaflets, material provided by a third-party (e.g. documents written in any standard programme such as *Word*), other material such as a calculator, graph plotter or other mathematical software (eg. worksheets written in, say, *Maple* or *Mathematica*). Any cross-referencing between the modules is supervised by a sophisticated piece of software which is called the Courseware Manager (CWM) that keeps track of where the student is and provides access to available resources using an internal search engine. Modules have been authored on both PC and Macintosh platforms and illustrative layouts for pages of a learning units are shown in figures 4.1 and 4.2, which show the general commonality of features. A full list of existing modules is given in table 4.1.

Note that the application appears platform independent. The most important text appears on the left-hand side of the screen; subsidiary text appears on the right to allow for any additional information that appears as a 'pop-up' covering it. Navigation tools are kept to a minimum in the bottom right hand corner. Included in the navigation tools is access to a search engine that also provides links to graphical, biographical other facilities developed as part of the system.

Mathwise is however more than a collection of individual modules as its functionality is increased dramatically by the inclusion of a Courseware Manager (CWM) that enables *Mathwise* to be configured to a particular user's requirements, controls the student use of the system, manages existing resources and allows additional resources to be included. A further advantage is that internally the development of many learning units is based on the creation of 'leaflets' that contain key material that can be accessed directly by the student. Thus leaflets act to provide a general reference material globally throughout *Mathwise*. Being external, they are accessible to and may be called from learning units from any other modules under the control of the CWM. It is through this means that the learner may cross-reference material from outside the unit being studied. The learning units and leaflets from differential calculus modules on 'Concepts of Differentiation', 'Rules of Differentiation' and 'Maxima and Minima' are shown below. Much of the information in the leaflets in 'Concepts of Differentiation' is required in subsequent modules and can be readily accessed using the CWM to make appropriate searches.

Table 4.1. List of Existing *Mathwise* Modules

PC Modules	Mac Modules
Real Numbers and Algebra	Real Numbers and Algebra
Graphs of Functions	Graphs of functions
Inequalities	Inequalities
Linear Equations and Matrices	Trig Equations
Polynomials, Powers and Logs	Concepts in Differentiation
Trig Functions	Rules of Differentiation
Standard curves and conics	Minima and maxima
Euclidean Geometry	Complex Analysis II
Concepts of Differentiation	Solution of liner equations
Rules of Differentiation	Fourier Series
Minima and maxima	First order ODEs
Concepts in Integration	Eigenvalues and Oscillations
Techniques of Integration	Algebraic Eigenvalue Problem
Complex Numbers	
Sequences and Series	Differential Equations
Solution of Linear Equations	Discrete Mathematics
Basic Vector Algebra	Elec. & Electronic Eng.
Linear Programming	Business Studies
Multiple Integrals	Data Fitting
Higher Order DEs	Mathematical Modelling
Numerical Solution of DEs	
Algebraic Structures	Mathematical Biographies
Astronomy	
Differential Equations	
Earth Sciences	
Biology	
Discrete Mathematics	
Mechanics	
Coding Theory	
Mathematical Modelling in Sport	
Mathematical Biographies	

Figure 4.1: Sample PC Modules Pages

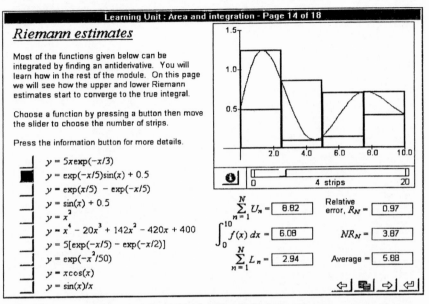

Figure 4.2: Sample Mac Modules Pages

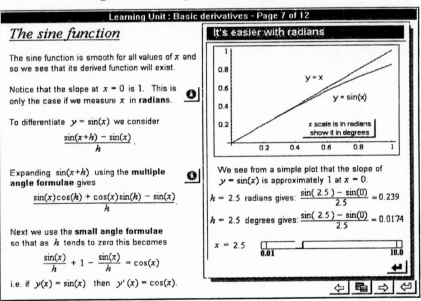

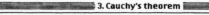

3. Cauchy's theorem

The primitive of a Complex function

Drag the mouse to make a curve to integrate along.

The function being integrated on the right is

$$f(z) = z^2 + z + 1$$

and its **primitive** is any function of the form

$$F(z) = z^3/3 + z^2/2 + z + C$$

for a constant C.
If you choose a point where $F(z) = 0$ as a starting point then the value of the integral will match the value of the primitive F shown.

f(z) = | 8.251-0.622i |

F(z) = | 8.469-0.937i | (with C = 0)

$$f(z) = z^2 + z + 1 \quad \square$$

$$f(z) = z\sin(z) \quad \square$$

$$f(z) = z\exp(z^2) \quad \square$$

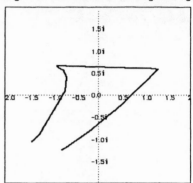

Coordinates: 2.241-0.113i
Starting point: 1.291+0.582i
Integral: -2.443-2.632i

| Clear picture |

Page 8 of 10

Normal Modes

Vibrating Masses Solution

Let us consider the equations of motion of two masses linked by springs as shown below

| Start |

The equations of motion are
$$my'' = k(-2y + z)$$
$$mz'' = k(y - 2z) + 4kL$$

We write these equations as

$$Y'' = BY + c$$ 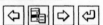

Eliminating c by setting Z=Y+r gives

$$Z'' = BZ$$ 

The eigenvalues and eigenvectors of B determine
    (i) the normal modes
and
    (ii) the solution Y(t).

Examples

Normal Modes.

The eigenvalues of $\begin{bmatrix} -2 & 1 \\ 1 & -2 \end{bmatrix}$ are -1 & -3.

so the eigenvalues of $B = \frac{k}{m}\begin{bmatrix} -2 & 1 \\ 1 & -2 \end{bmatrix}$ are $-\frac{k}{m}$ & $-3\frac{k}{m}$.

The corresponding eigenvectors are $\begin{bmatrix} 1 \\ 1 \end{bmatrix}$ and $\begin{bmatrix} 1 \\ -1 \end{bmatrix}$

Normal Mode 1:

$$Z_1(t) = A\begin{bmatrix} 1 \\ 1 \end{bmatrix}\cos(\sqrt{\tfrac{k}{m}}t + w)$$ | Mode 1 |

The masses vibrate in phase with the same amplitude

Normal Mode 2:

$$Z_2(t) = A\begin{bmatrix} 1 \\ -1 \end{bmatrix}\cos(\sqrt{\tfrac{3k}{m}}t + w)$$ | Mode 2 |

The masses vibrate out of phase with the same amplitude, this motion is at a higher frequency than normal mode 1.

Page 6 of 7

Table 4.3: Learning Units and Leaflets for "Concepts of Differentiation"

M4BASE Concepts of Differentiation	
Learning Units	Leaflets
Introduction	Slope of a line
The idea of slope	Rate of Change
Basic Definition	Basic Definition
The derived function	Limits
Polynomial Functions	Notation
Log Functions	Standard derivatives
Trig Functions	Functional form
Notation	Higher order derivatives

Table 4.4: Learning Units and Leaflets for "Techniques of Differentiation"

M4DERV Rules of Differentiation	
Learning Units	Leaflets
Sum rule	Definition
Product rule	Basic derivatives
Quotient rule	Chain rule examples
Chain rule	Chain rule exercises
Higher order derivatives	$d/dx(\text{root } x)$
	Derivative of $\cosh(x)$
	Derivative of $\sinh(x)$
	.
	.
	Derivative of inverse $\cosh(x)$
	.

**Table 4.5: Learning Units and Leaflets for
"Applications of Differentiation: Maxima and Minima"**

M4DERV Rules of Differentiation	
Learning Units	Leaflets
Max/min theory	First derivative test
Max/min test	Global Max and min
Max/min practise	Increasing & Decreasing func-
Curve Sketching	tions
Series	L'Hopital's Rule
L'Hopital's Rule	Local Max and Min
	Mean Value Theorem
	Second derivative test
	Taylor series
	Examples of max/min problems

4.2 Assessment in *Mathwise* Modules

Experience has shown that even when students have taken a diagnostic test, if they are given some credit for any "remedial" work suggested it is more likely that they will follow it through. A clear way to ensure that they do such work is to provide some form of assessment. However student-specific requirements, such as those produced through diagnostic testing, can often prove difficult but they can be overcome using the built in assessment mechanisms of *Mathwise*. Within each module there is considerable formative assessment to help students monitor their progress and understanding. However, many *Mathwise* modules also have a more formal Assessment Section. This test can be taken in a number of different modes that are designed to be configured by individual instructors. These modes are:

•   **Examination mode** — no visible marking on screen ;

•   **Practice mode** — marking at the end of each question which is good for monitoring good students;

•   **Learning mode** — ticks and crosses at the end of each part of a question which is good for monitoring moderate students;

- **Help mode** — with ticks and crosses at the end of each part and the chance to reveal answers if you are stuck and this mode is useful in helping to build student confidence.

It is possible to choose to take questions on just one learning unit or rearrange the order in which questions are presented to fit in with the delivery. The students can view their own progress as they take a test, have access to formulae and can save their test so that it can be completed later. Examples of formal test pages are shown in figures 4.3 and 4.4.

Figure 4.3: An Assessed Question from the *Mathwise* Module 'Maxima and Minima'

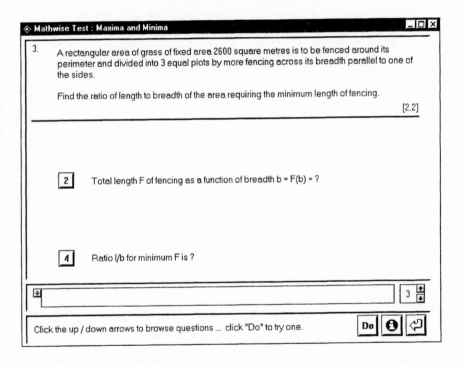

Figure 4.4: The *Mathwise* Input Tool

◈ Mathwise Test : Maxima and Minima   _|□| x|

4.
   Find the value of x at which the curve $y = \dfrac{x^2 - 1}{x - 2}$

   has a vertical asymptote. Determine the values of x at which the curve crosses the x-axis and
   the value of y where it crosses th| **Input Tool** .... X| Determine
   the y-coordinates of these points

                                                                                         [1,1,1,1,2,2]

   | 1 | x value for vertical asym $\dfrac{\left(x^2 - 1\right)}{(x - \ ?)}$?

   | 2 |

   | 3 |

   | 4 |

[+] | 6 | |(x^2-1)/(x-|

                                    About   Cancel   OK

[+] | 8 |

|↓| 4 |↕|

Enter your answer or press Cancel (or enter null) to exit. |End| 🔴 |↵|

The figures also illustrate some features of the Mathwise Test. Firstly it is possible to browse through all the questions (like you would during a traditional examination paper). When a student decides to answer a particular question then they simply select it. Secondly, each *Mathwise* question contains KEY-PARTS, a KEY-PART may sub-divided into have SUB-PARTS. If a student finds a KEY-PART too difficult to answer then it is possible to reveal the SUB-PARTS. A further feature included in the *Mathwise* test is the inclusion of an 'Input Tool' that helps the student insert mathematical formulae from the keyboard in response to test questions. When a student enters an answer in one-line input in mathematical notation the Input Tool re-displays the result in a two-dimensional display; input builds up dynamically, giving a clearer representation of how the computer interprets the keyboard input (see Figure 4.4).

5. A Case Study: Diagnostic Testing and Support at Keele University

At Keele University the introduction of computer-based diagnostic testing was seen to:

- Encourage students to be self-critical;
- Allow students to obtain immediate and individual feedback;
- Give students their own learning profile;
- Allow students to direct their own additional studies.

A decision was made to introduce a simple diagnostic test based on the Nottingham shell for all students entering principal Mathematics to investigate their mathematical skills. The test was implemented by the authors and comprises twenty questions selected from a pool of questions to be answered in 40 minutes; selected pseudo-randomly to ensure little chance of test duplication but maintain an adequate range of topic coverage. A prompt line giving the student advice on the operation of the test is active throughout and an option is provided for the student to view an animated introduction to the test. Typical questions are shown in figure 5.1 and 5.2, with the question statement on the left-hand side and four possible answers, randomly positioned, on the right. Questions can be selected and answered in any order. Furthermore, students can usefully repeat the test at future times to gauge improvements in their abilities.

In diagnostic mode the objective of the test is to assess the student's performance against a selected set of ten key topics. Diagnostic questions can address a single topic or more elaborate questions will require two or more of the key topics to be deployed. In the latter case, the compilation of the alternative (incorrect) answers can be used to delineate the source of the error. With careful construction of alternative answers an incorrect response can used as part of the diagnostic element based on the choice of incorrect answer. In this way students can be rewarded for partially correct answers by identifying the key skills needed to obtain the answer. Globally, each student response is judged in terms of the skills required to gain a correct answer, a question by question performance a total (percentage) score calculated weighted to minimise 'guesses' (+3, for a correct answer and -1 for an incorrect answer). More importantly, an individual profile is presented to the student such as shown in figure 5.3 as the composition of their correct and incorrect answers.

Figure 5.1: Testing Integration

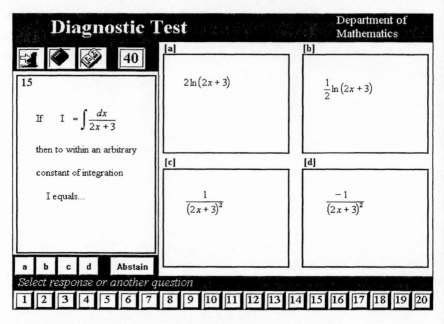

Figure 5.2: Testing Polar Co-ordinates

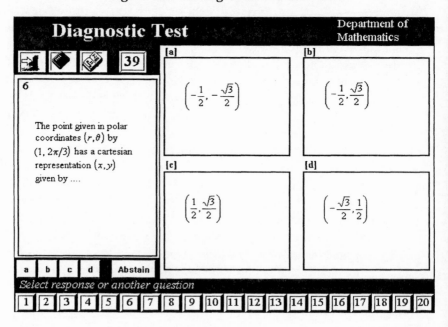

Figures 5.3: Student Diagnostic Report

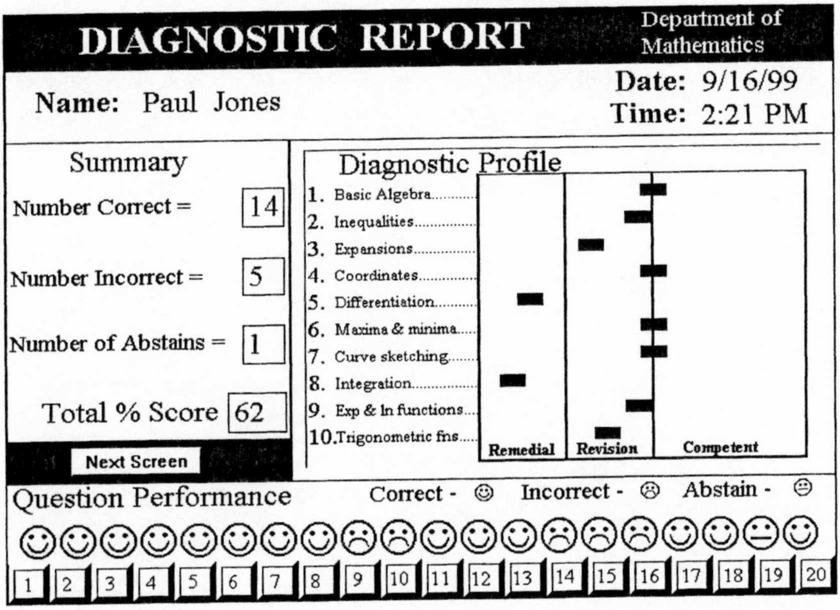

To direct the student towards the appropriate support material a final screen is presented to the student, as shown in figure 5.4, that interprets the student profile in terms of the topic sections in *Mathwise* that should be studied.

Examination of the raw data files provide very detailed information on total student performances that can be readily used as a reference for module revisions or for directing supplementary resources. Further, global information on each question can be readily extracted including the most common incorrect answers.

Although our primary aim is to help students to achieve their potential the gathering of national information regarding the mathematical abilities of students in higher education is also possible. Such information will help long term planning in higher education and also provide positive feedback to secondary education in general.

A diagnostic and support system has been operating since 1995/96 and figure 5.5 illustrates results of profile skills for the whole student cohort in four successive years. The wide discrepancy, year by year, indicates that simply selecting all students and providing common remedial courses will not be suitable. Indeed the variation over time suggests that a different set of support material would ideally be needed each year. It seems appropriate, therefore, to look at the microscopic scale and try to focus on individual students and attempt to assign each student suitable support material. Providing individualised programmes of study using computer based self-study programmes, such as *Mathwise*, may be a solution to this problem.

Figure 5.4: Suggested Action for Student

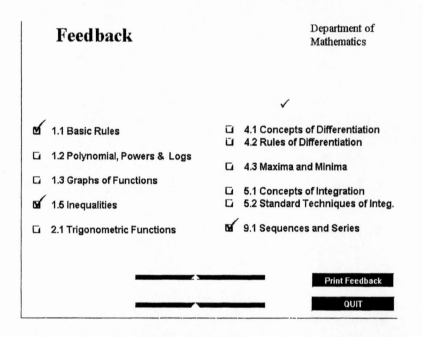

The results of the diagnostic test between 1996-98 were sufficiently encouraging that it was decided to integrate the process diagnosis-support into the first year programme. Follow the diagnostic test students were allocated to one of the three *Mathwise* modules in the Differential Calculus. After completion students were expected to take the associated test and the recorded score was incorporated into their module grade. The additional overheads in students submitting work on a variety of modules was minimal since the diagnostics marks, required module and assessment were all collated automatically. The response from students has been exceptional in that they have suggested that we extend the process to consider Integration in more detail.

Conclusions

Existing technology and courseware is available to help detect areas of mathematical weakness at individual student level. Furthermore, there is a sufficiently large base of computer assisted teaching material, in particular *Mathwise,* to provide links with support material in a semi-automated way. Although discussions with course tutorial support staff are vital, the computer-based profiles provide a pro-active mechanism for the early identification of student weaknesses. The basis of this paradigm is dependent on the development of study skills by individual students and the inclusion of both summative and formative assessment can help re-enforce this. The same software can also be used to gather information on the cohort as a whole and also to track the performance of students on a year-by-year basis.

**Figure 5.5: Averages of Student Profiles from
Keele University 1996-99**

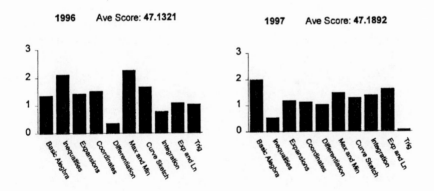

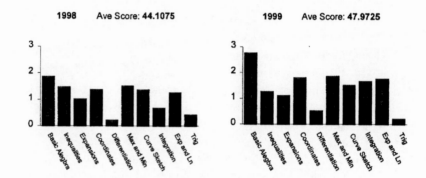

Although the primary aim of the test is to help students to achieve their potential by detecting possible weaknesses, the gathering of national information regarding the mathematical abilities of all students is possible. Such information will help long term planning in HE and also provide positive feedback to pre-university education in general.

References

Beevers C. E., Foster M. G., McGuire G. R. and Renshaw S. H. (1992) *Some Problems of Mathematical Cal.* Int. J. Comp Educ. 18 pp. 119-125.

Beevers C. E., Maciocia A, Prince A.R., and Scott T.D. (1996) *Pooling Mathematical Resources* Active Learning 5 pp 41-42.

Brydges S & Hibberd S, (1994) *Construction and Implementation of a Computer-Based Diagnostic Test.* CTI Maths and Stats. 5/3 pp9-13.

CTI Report, (1996) *Computer Based Assessment in Mathematics.* CTI Maths & Stats, 5/4 pp 26-29.

Harding R. D., Lay S.W., Moule H. and Quinney D.A. (1995) *Multi-media interactivity in mathematics courseware: The Mathematics experience of the Renaissance Project.* Computers in Educ. Vol 24. No 1, pp 1-23.

Harding R. D. and Quinney D.A. (1996) *Trends in Calculus Courseware.* Proceedings of Seventh Annual International Conference on Technology in Collegiate Mathematics, Addison-Wesley, pp 194-198.

Harding R. D. and Quinney D.A. (1997) *Mathwise: Basic Mathematics and Calculus Courseware.* Proceedings of the Eighth Annual International

Conference on Technology in Collegiate Mathematics. Addison-Wesley, pp181-185.

Hibberd S, (1996) *The Mathematical Assessment of Students Entering University Engineering Courses.* Studies in Educational Evaluation 22 pp 375-384.

Hirst K. (1997) *Changes in A-Level mathematics from 1996.* University of Southampton.

Kitchen A. (1996) *A-Level Mathematics isn't what it used to be; or is it?* Mathematics Today. 32/5 pp. 87-90.

Lawson D (1967) *What can we expect of A-level mathematics students?* Teaching Mathematics and its Applications 16/4. pp 151-156.

London Mathematical Society. (1995) *Tackling the Mathematics Problem* Report

SECTION 3
ATTITUDE

7

USING "SEMINARING" TO ACTIVELY
ENGAGE STUDENTS IN A CALCULUS CLASS

Philip A. DeMarois

Abstract

A seminar brings together a group of learners who have done some advance preparation to discuss their understandings of the topic. This paper includes a description of the implementation and evaluation of the seminar as a cornerstone of a college calculus class.

Introduction

In recent years, mathematics educators have increased the use of active learning strategies in mathematics classrooms as endorsed by Meyers and Jones (1993) who write: "A growing body of research today points to active learning strategies — in which students talk and listen, read, write, and reflect as they become directly involved in the instructional process — as a way to better engage students, cultivate critical thinking, and improve the overall quality of teaching and learning" (front cover). *Crossroads In Mathematics: Standards for Introductory Mathematics Before Calculus* (AMATYC, 1995, p.11) includes "Communication: Students will acquire the ability to read, write, listen to, and speak mathematics" as a Standard for Intellectual Development. Under Standards for Pedagogy, this document lists: "Interactive and Collaborative Learning: Mathematics faculty will foster interactive learning through student writing, reading, speaking, collaborative activities so that students can learn to work effectively in groups and communicate about mathematics orally and in writing" (ibid., p.16).

Meyers and Jones (1993) suggest that active learning makes two assumptions: "(1) that learning by nature is an active endeavor and (2) that different people learn in different ways" (p. xi). They suggest that the key components of active learning include listening and talking to force clarity of thinking, writing, reading, and reflection.

This paper discusses a technique called "seminaring" that is designed to promote each of these four components of active learning. Before listening to a lecture or doing problems, students concentrate on reading the text, writing answers to conceptual questions, meeting in groups to discuss their answers to the questions, and agreeing on a group answer to each of the questions. The details of this approach are the key points of this paper.

Theoretical Framework

Each component of active learning involves different ways of thinking and helps students create different mental structures. It is the development of these mental structures that leads to understanding. Hiebert and Carpenter (1992) define understanding as follows:

> A mathematical idea or procedure or fact is understood if it is part of an internal network. More specifically, the mathematics is understood if its mental representation is part of a network of representations. The number and the strength of the connections determine the degree of understanding. A mathematical idea, procedure, or fact is understood thoroughly if it is linked to existing networks with stronger and more numerous connections. (p. 67)

Active learning techniques are designed to increase the number of connections students develop among mathematical ideas. In the process, it is hoped that their understanding is relational, rather than instrumental, as described by Skemp (1976). Relational understanding is demonstrated by knowing what to do and why while instrumental understanding involves a multiplicity of rules rather than fewer principles of more general application. Skemp writes: "Instrumental understanding necessitates memorizing which problems a method works for and which not, and also learning a different method for each new class of problems" (ibid., p. 23). The advantages of relational understanding include improved adaptability to new tasks and less dependence on memory. Relational understanding promotes the building of effective connections internally. Active learning techniques are designed to promote just such understanding.

Hiebert and Lefevre (1986) make a similar distinction when they contrast conceptual knowledge which is "knowledge that is rich in relationships"(p. 3) with procedural knowledge which is "composed of the formal language, or symbol representation system...[and] the algorithms, or rules, for completing mathematical tasks" (ibid., p. 6). They go on to assert that procedural knowledge is meaningful only if it is linked to a conceptual base. One focus of active learning strategies is to develop a firm understanding of concepts. In order to promote conceptual knowledge, Hiebert and Carpenter discuss what must be done instructionally: "Teaching environments should be designed to help students build internal representations of procedures that become part of larger conceptual networks before encouraging the repeated practice of procedures" (1992, p. 79). The use of a "seminaring" technique builds just such an environment in which students concentrate on the development of a conceptual network before focusing on problems.

What "Seminaring" Is

The seminar brings together a group of learners who have done some advance preparation, including reading, thinking about (reflection), writing about, and solving problems related to a specific topic. Prior to a seminar, students are given an assignment that includes reading and problems. They receive a series of conceptual questions on the assignment to write responses to prior to the next class session. During the seminar time in class permanently assigned groups meet to discuss their written responses. Students refine their responses based on the group's discussion.

Implementation of the Seminar in Calculus

The technique of "seminaring" was the primary mode of operation in two sections of Calculus I during the 1996-1997 academic year. On the first day of class, students were given a policy statement explaining the seminaring concept. Emphasis was placed on the goal of actively involving the students in their own learning. Critical to the process would be their willingness to read their textbook carefully and to write thoughtful answers to conceptual questions based on their reading. Students were informed that they would be assigned to permanent groups of size 3-4 for the purpose of discussing their written answers to the seminar questions. These groups also served as study groups in which students helped one another with homework problems and with exam preparation. The groups were assigned based on student input and on information collected by the instructor in order to assure each group had an appropriate mix of abilities. Space does not permit more detail here on group formation.

Each day, students were given a handout detailing the assignment for the next class. This handout contained seminar questions on the reading and a list of problems to try to test understanding. For example, on a section that introduces the concept of derivative, the seminar questions were as follows:

1. The derivative is a function in which the input is a function and the output is a new function. Your friend, Gordo, is confused about this. Describe to Gordo the meaning of the output function in terms of the input function.
2. Under what conditions will the derivative of a function fail to exist at a specific point?
3. Discuss the relationship between a function being differentiable at a point and a function being continuous at a point.

Each set of seminar questions included three questions asking students to reflect on their comfort level with the material, as follows:

1. What questions do you have after studying Section n?
2. Were you able to do the assigned problems? Which was the easiest? Why? Which was the hardest? Why?
3. What questions do you have about the assigned problems?

Once in groups students were asked to compare and contrast their answers to the conceptual questions. Once they had reached agreement, one person was designated as the recorder and wrote a group response to each of the questions. Following this, each group could then discuss homework problems, as time permitted.

At the end of each seminar, the instructor collected the individual seminar papers and the group seminar response. The group response was graded based on how well the answers indicated understanding of the concepts. Individual papers were checked for completeness. Individual student responses to the "reflection" questions were noted.

Seminars occurred in approximately ninety percent of the class meetings. Lecturing was restricted to those concepts and problems that proved difficult to several groups. While students were in groups, the instructor circulated answering individual group questions and supplying mini-lectures, as needed. A key role of the instructor after a seminar session included summarizing the key points of the seminar and helping students build connections among the various concepts.

Student Evaluation of the Seminars

Students completed an extensive, specially-prepared course evaluation at the end of the semester. Included were statements that the student responded to on a Likert scale. In addition, students were encouraged to write comments on each major aspect of the course. The following collection of charts indicated student responses to various questions that relate to the use of the seminar technique.

Question 5: To what degree do you think **this course has improved your ability to read and understand a mathematics text?** (1 = not at all; 2 = a little; 3 = somewhat; 4 = a good bit; 5 = very much)

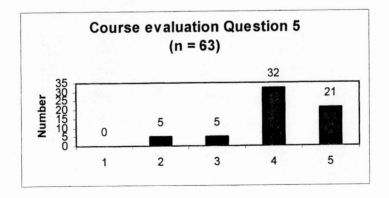

Notice that approximately 84 percent of the students indicated that their ability to read and understand a mathematics text had improved a good bit or very much.

Question 6: How would you rate your **ability to write about mathematical concepts at the beginning of the semester?** (1 = very poor; 2 = somewhat poor; 3 = fair; 4 = somewhat good; 5 = very good)

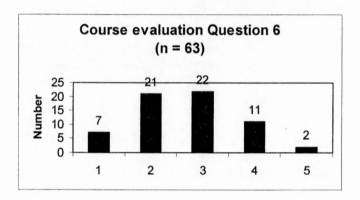

Question 7: To what degree do you think **this course has improved your ability to write about mathematical concepts?** (1 = not at all; 2 = a little; 3 = somewhat; 4 = a good bit; 5 = very much)

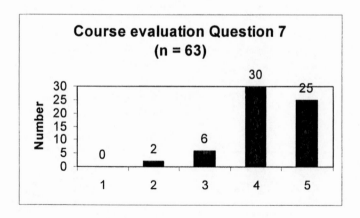

Again 87 percent of those responding suggest that their ability to write about mathematical concepts had improved a good bit or very much.

Question 11: How would you rate your **ability to work effectively in a group at the beginning of the semester?** (1 = very poor; 2 = somewhat poor; 3 = fair; 4 = somewhat good; 5 = very good)

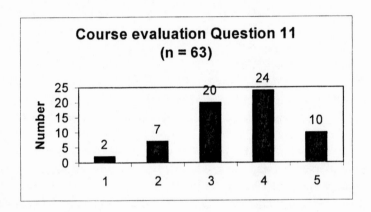

Question 12: To what degree do you think **this course has improved your ability to work effectively in a group?** (1 = not at all; 2 = a little; 3 = somewhat; 4 = a good bit; 5 = very much)

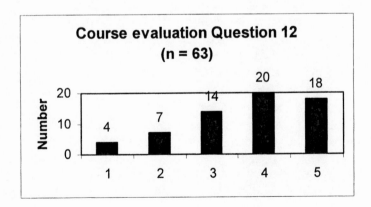

While approximately 86 percent said that they were at least fair at working in groups at the beginning of the semester, approximately 83 percent indicated that this ability improved at least somewhat as a result of the course.

Students were asked to provide written comments on each aspect of the course on this evaluation. A few of the comments on seminars follow:

- They are a very different approach to mathematics from what I am used to. I feel they are a much better way of learning math. You're not just doing every other odd problem. You are learning to understand what you are actually doing.
- They were a good learning aid. It helped us learn each concept, and a good way to let you know of any questions we might have. A good push to get us to study each section to understand that section's concept.
- They really helped. By having to write about a concept, you really had to understand what you just read. Also, by writing about a concept, it helped me remember it.
- I enjoyed the seminar papers because they forced me to understand the concepts and be able to explain them in my own words and with the use of examples.
- Extremely helpful in learning and understanding the various sections of the chapters. Helps create a more intuitive understanding of the concepts discussed.
- The seminar papers are probably the best idea of the class. Compared to conventional methods of teaching, for me, this way is much more useful. I started doing something like a seminar in my chemistry class and it helped me understand there also. The seminars were a useful extension of the papers. It is also a whole lot more interesting than listening to a teacher lecture.
- I loved it. I actually got to learn concepts. From these, the homework was easy. In other courses I learned problems and when the problem would change I needed to be spoon fed again. But the seminar papers concentrate on what I should know to attack any problem.
- By participating in the seminars I was able to depend on myself more for learning. By reading each section carefully, answering questions, and doing some problems I think I was able to grasp the concepts a lot better than if someone were to tell me how to do it. I think seminars are a wonderful way to learn math. Of course they have to be accompanied

by group discussions and teacher guidance. After I got used to doing them, I found that I learn more by reading the book, answering questions about what I read, and then discussing it with others.

• Seminars have helped me to form new independent study habits and self-reliance. But, most of all, they have forced me to take the time to thoroughly learn and understand a concept as best I can before class discussion.

The above testimonials indicate that the goals of active learning appear to be well served by the seminar technique.

Pros, Cons, and Modifications

The pros of using the seminar approach are numerous. These include helping students become active and independent learners, improved group skills, improved ability to read and understand a mathematics textbook (ability to use resources), more focus on conceptual (relational) understanding as opposed to procedural (instrumental) understanding. Focusing on the confusing concepts rather than on all concepts uses class time more efficiently.

Some cons to the approach include time commitment for both instructor and student, grading time, caution so as not to excessively punish poor writers, and time required preparing the seminar questions. However, from student evaluations and instructor observation of student understanding, it appears that these problems are more than outweighed by the positive results.

Note that students need guidance on both how to read mathematics text and clear instructions on how to write responses. It should be clear whom they are writing to? Common seminar questions asked students to explain a concept to a confused friend. It is important for the instructor to be clear about the instructional objective. It is helpful to share the potential benefits and the theory behind this active learning approach with students.

Since the experience with seminars in these calculus classes, the instructor has gone on to incorporate the technique into other classes. One change in procedure has been instituted to give the instructor an instant measure of how well students understood the reading. Each seminar assignment includes the following task:

Place an X on the continuum below indicating how well you understood the section.

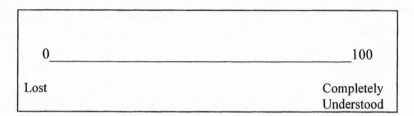

When students are directed to move into groups, they are first asked to place an X on the same continuum on the board matching where they located their X on the seminar paper. The instructor gets a quick summary of how well the class understood the assignment. At the end of the seminar or the class, students can mark the continuum again to indicate their understanding at that point in time. This technique is extremely helpful to the instructor in orchestrating class discussion.

Conclusion

The use of the seminar truly transformed the class into an active learning environment. Students actually read the text. They write about mathematics in a thoughtful way. They come to class prepared to discuss concepts as opposed to listen passively to a lecture or watch a problem being solved. They write and talk mathematics. They initially may resist this approach as "different" and not what they are "used to." Early seminar papers are quite poor, as this is something students have never done before. With careful constructive criticism, they improve as they do more. Their confidence in their ability to read and understand mathematics independently grows.

References

American Mathematical Association of Two-Year Colleges (AMATYC) (1995). *Crossroads in Mathematics: Standards for Introductory College Mathematics Before Calculus*. Memphis, TN: AMATYC.
Hiebert, J. & Carpenter, T. P. (1992). Learning and Teaching with Understanding. In Grouws, D.A. (Ed), Handbook of Research on

Mathematics Teaching and Learning. New York: Macmillan Publishing Company. pp. 65-97.

Hiebert, J. & Lefevre, P. (1986). Procedural and Conceptual Knowledge. In Hiebert, J. (Ed.), *Conceptual and Procedural Knowledge: The Case of Mathematics*. Hillsdale, NJ: Erlbaum. pp.1-27.

Meyers, C. & Jones, T. B. (1993). *Promoting Active Learning: Strategies for the College Classroom*. San Francisco, CA: Jossey-Bass Publishers.

Skemp, R .R. (1976). Relational Understanding and Instrumental Understanding. *Mathematics Teaching 77*. pp. 20-26.

8

CHANGING ATTITUDES AND PERSPECTIVES OF PRE-SERVICE TEACHERS ABOUT MATHEMATICS

Leonard J. Lipkin

Pre-service elementary and middle school teachers, as well as in-service teachers who are updating certification, enroll in courses in the Department of Mathematics and Statistics. The course that I will focus on here is a required one, normally taken after two semesters of college mathematics. The previous courses vary considerably among the students in content and in the time that has elapsed since they took the courses. They might have taken "Intermediate Algebra," "College Mathematics," "Finite Mathematics," "College Algebra," or some other course. Despite recent advances in "mathematics reform," virtually all of the students took at least their early mathematics courses prior to the reform movement and the NCTM Standards. When these students become teachers they will be expected to incorporate new approaches to the curriculum in their classes. We must address these issues: (1) Their attitudes toward mathematics; (2) Their perceptions about the meaning and uses of mathematics; (3) Their understanding of content and relationships among branches of mathematics; and, (4) Their ability to communicate mathematics.

These issues are related to each other. For the purpose of this short paper, I will describe an approach to them and some of the results.

An inventory is taken at the beginning of class. In addition to usual factual items, students are asked what parts of mathematics they like, what parts they don't like, what they see as the reasons for studying mathematics, what they think it takes to be a good mathematics teacher, and then: "How do you feel about mathematics?"; "How do you want your students to feel about mathematics?" These questions are important so that the instructor understands the students and, equally important, so that the students know how others in the class feel. Typical responses are: "I can go along with the need to learn the basics, but that's all."; "I find it very frustrating but sometimes it's o.k."; "I've never been good in math; I don't like it."; "I kind of like it, but it's not my favorite subject: I've always done well in it."; "I get nauseous when I think of mathematics."; "I hate mathematics!" Very few respond that in order to be a good mathematics teacher you need to know something about the subject! Despite the overwhelming majority who fear, dislike, or don't really understand what mathematics is all about, all of these prospective or current teachers want their students to feel confident about mathematics, enjoy it, and be successful. How can they make that happen?

The goals of the course are: (1) Show the students that mathematics is fun and that there are many interesting, and even astounding results; try to get reactions like "that's neat!"; (2) Reduce anxiety about mathematics; (3) Show the students that there is much more to mathematics than skills, formulas, rote memorization, and arithmetic (or algebra); it is a process, a way of thinking, and a way of exploring the power of the mind; (4) Work toward a deeper understanding of standard topics, and show relationships among different branches of mathematics; (5) Understand the need for communicating mathematics; and, (6) Change the students from passive learners into confident teachers of mathematics.

Those are certainly lofty goals. We work toward those goals using a variety of techniques that I will describe below. When we began to make serious changes in the course (in the late 1980s), most textbooks were very traditional. Today the books have changed, and many contain a wealth of interesting material and activities. Nevertheless, the selection of topics and particularly the classroom dynamic are what account for the modest successes that have been enjoyed. The instructor must be ready to follow routes and explore topics that arise naturally in the classroom.

We assign small, outside projects to be done by groups of three students. These are intended to get the students thinking about the sizes of numbers, units, conversions, measurement, how to find information in the library or internet, and how to write a mathematical report. For example, "how many dollar bills does it take to make a stack as high as the Washington Monument?"; "How many teaspoons of sand does it take to get one million grains?"; "If you have one million drops of water could you drink it, bathe in it or swim in it?" (Courtesy of Jim Wilson); "How many teaspoons are there in a cubic light year?" (Courtesy of Clemson University) A valuable project that connects the physical and mathematical worlds is the construction (or drawing) of a scale model of our solar system. Students will see that many textbook renderings are not at all accurate. Another project begins with a classroom discussion: It is a rainy afternoon, and you decide to rearrange the four books on your bookshelf. How many arrangements are there? How many are there with 10 books or with n books? This is standard fare, and we arrive at the result $n!$ as usual. Now comes the question: suppose you have 20 books and it takes (on the average) 15 seconds to complete each rearrangement. How long will it take to complete them all? The answer is astounding! It takes longer than the time that has passed since the Big Bang! *That* gets their attention. Students can now ask questions like: If a Pilgrim landing at Plymouth Rock began to go through all rearrangements of n books on a shelf, what value of n would allow the Pilgrim to finish by the year 2000? Questions such as this one help the prospective teachers see how to create examples for their own students. These projects also help reduce anxiety because the students can certainly complete them, and they learn to think about the meaning of the numbers that arise as "answers" to textbook problems. Finally, they must write a report using complete sentences and meaningful paragraphs.

An activity that has proved to be very useful in improving students' reasoning, while allowing them to see that they can reach levels that they did not think possible, and have a lot of fun, is to use a series of logic puzzles that build in sophistication. Excellent sources are the books *The Lady or the Tiger* and *What is the Name of this Book* by Raymond Smullyan. For example, begin with the scenario:

All the inhabitants of a land are either knights or knaves. Knights always tell the truth, and knaves always lie. Walking along in this land, you come upon three people — A, B, and C. You ask A, "Are you a knight or a knave?" The person answers, but mumbles so badly that he could not be understood. So you ask B, "What did A say?" B replies, "A said that he is a knave." At this point the third man, C, says, "Don't believe B; he is lying!" The question is, what are B and C? Can you tell what kind of person A is?

This is an old problem. But, what is significant here is that students generally think it is impossible to solve the problem, or at least they have no idea how to begin. After some hints and discussion, we collectively solve the problem. In typical books, the topic is now dropped. This is the shortcoming. Our technique is this: Each week hand out at least three more of these following, for example, the second book of Smullyan above. We require each student to hand in a clearly reasoned solution to each of the problems. The problems build in sophistication, and as the term progresses, the students find that they become adept at solving these problems. Their reasoning skills improve, and they begin to have fun working on the problems. They see that they are able to accomplish something that they never thought possible. They find that writing clear solutions to problems is a challenge and a necessity. And, after just a week or two, most students report that their families or roommates have become so interested that they look forward to the new set each week. Imagine! A family sitting down after dinner to work mathematics problems!

Also each week there are problems of a more typical mathematical nature. For example, how many zeros are there at the end of 100! ? This question shows them that calculators do not provide the answer to everything. The students must now rely on their knowledge of factoring to arrive at the solution. Many of the students who intern in later semesters (or who are current in-service teachers) use some of the problem sheets from this class with their own students. They report that the children really get excited about working on them. Just as with the logic problems, it is important to give problems regularly throughout the course so that students can build their skills.

A central technique in teaching this course is to take every opportunity to show students that pursuing questions beyond their simplest form may lead to very interesting connections among the branches of mathematics. For example, it is standard to observe that the theorem of Pythagoras shows that a segment can be constructed whose length is $\sqrt{2}$ by constructing a right triangle with legs each 1 unit long. A proof is usually suggested to show that $\sqrt{2}$ is irrational. A typical text will drop the subject there, or perhaps it will inquire about $\sqrt{3}$ and then drop it. The message to the student is once again: one more thing to memorize and another dead end. But instead, more can be done. We can ask about constructing right triangles whose legs are integers. This leads the students to think about what whole numbers are the sums of squares of integers. From this geometry discussion arises a number theory question leading to the Fermat "sum of two squares" theorem. The statement of this theorem about the primes that are sums of squares can be easily discovered by the students. (Naturally, we don't consider the proof at this level.) They in turn see how they can use this method in their own classrooms. This exploration leads naturally to discussing primes and discovering other patterns. At this point students can see that these two branches of mathematics are intimately connected. For the middle school teachers, we can discuss Pythagorean triples, discover more patterns, and again connect geometry with number theory.

Let us digress briefly to talk about the training of middle school teachers. Because of a serious shortage of teachers who can teach mathematics at this level, school districts look for ways to train elementary school teachers and "second career" people to do the job. The ideas that we discussed above can be upgraded for this group. The study of Pythagorean triples is a perfect example of what can be done. (What follows is also very good for high school teachers.) First we ask students for examples of right triangles whose side lengths are whole numbers. After some observation and discussion it will become clear to them that the primitive triples are the building blocks for all of the Pythagorean triples. Next, the students can observe that many of their examples have the form $[n, (n^2-1)/2, (n^2+1)/2]$, n odd. It helps the students to actually calculate "a^2+b^2" to see that they get the correct result. They can also observe easily that each such triple is primitive and that if n is even this is not a triple of whole numbers. (This seemingly elementary exercise helps students practice reading formulas and think about divisibility.)

Some students will have found examples of primitive triples that are not given by the formula above. So they now have an interesting question to answer: Can we find all of the primitive Pythagorean triples? There is a standard derivation of Euclid's formula, the answer to this question. (See for example, *Excursions in Calculus*, by Robert Young, Mathematical Association of America.) The derivation replaces the original question by asking when a straight line through the point (-1, 0) intersects the unit circle in a point with rational coordinates. This derivation is very informative for prospective middle school (and high school) teachers because it shows how a variety of mathematics can be used together: geometry, equations of circles and lines, algebra, and number theory. (It also gives a chance to discuss the density of the rationals and to pose the challenge of showing that the circle $x^2+y^2=3$ contains no rational points. After the density discussion this seems to be truly amazing.) This helps to break down the common belief that mathematics comes in separate, unrelated packets named "arithmetic," "algebra," "geometry," and so forth.

The example above connecting geometry, number theory, and discovery shows that attention can and should be paid to mathematics as a creative, fun, intellectual endeavor. While it is important to use applications to motivate mathematical discussions, we need to emphasize by example that the human mind has a natural curiosity about the creative, "pure" aspects of mathematics. That example and the logical puzzles mentioned above show the students that mathematics is an intellectual process rather than a collection of dry facts and techniques.

Students can now use the web to expand their view of mathematics. Material is readily available relating mathematics to music, art, voting, surveys, and a host of other topics. The present and future teachers see still another resource so that they do not feel bound by the pages in their classroom textbooks. They can become active mathematics teachers rather than passive users of the texts for the grade level they teach.

The method of learning in these courses for pre-service and in-service teachers is highly interactive. The students present solutions orally, and free flowing discussions are the norm. Oral presentations make everyone immediately aware of the challenge of making clear mathematical statements and explanations of concepts. At first they are shy and reluctant to stand up and talk about mathematics. But encouragement (and require-ment!) usually will overcome fear. Writing is the other crucial side of communication. Even at this time in our renovation of mathematics instruction, few practicing teachers have much experience with writing sentences and explanations.

This brief summary shows the method used while covering material in a standard textbook. The focus becomes the process and the classroom dynamic rather than a collection of facts. In the end, of course, the students do collect the facts, but they are now more confident that they really own those facts. Most of the students are now more confident in their own abilities than they were when the course began. They have found that mathematics really is fun.

9

ARE WE AN "ALIEN?"

Jay M. Jahangiri
Jayne M. Kracker

"Alien" is the word that, unfortunately, can be applied to many (if not all) disciplines taught at our institutions from grade schools to higher education. Usually, the word "alien" is used to address one from another country — or another world — but it can also be used to define an institution that is "strange" or "unappealing" by virtue of its lack of access or involvement in its community. This would be the case as long as our "institution's products;" namely, "graduates," are incapable of or have difficulties in applying their academic teachings at the workplace.

It is no secret that there are competency gaps between the recent graduates of schools, colleges, and universities in the workplace. One of the problem areas that contributes to these competency gaps is "mathematics." A math competency gap not only exists among the newly hired graduates, but also among the experienced employees—those who have been on the job for a length of time.

We need to re-think and re-define the way mathematics is approached and taught in today's classrooms — from early school years to colleges.

No quick solution or magic remedy is effective for this problem. It needs two simultaneous approaches. One approach is long term and the other is focusing on "damage control." The purpose of the present article is to elaborate on *damage control*, but first we will give some information on the long-term solution. It is our hope that readers of this article will be inspired by our innovative ideas.

For the long-term solution, the focus needs to be on K-12 schools and institutions of higher learning, and especially math teachers.

"You can't teach what you don't know," but too many of our mathematics teachers may be doing just that: teaching what they don't know." (Wu, 1999).

"The awareness by organizations of mathematicians of the need to coordinate university mathematics departments across the land in order to upgrade the pre-service professional development of prospective teachers began to surface only in the past year." (Wu, 1999).

"Our teachers need and have to be life-long learners of new and innovative teaching ideas. An organized and nationwide in-service professional development is needed to mobilize our teachers. Because any improvement in education must start with improvement of the teachers already in the classroom, this topic is one of real urgency." (Wu, 1999).

"In the crudest terms, there are two kinds of in-service professional development: enrichment and remediation. The former is devoted to enlarging the mathematical knowledge of teachers who are already at ease with the mathematical demands in the classroom. The goal is to inspire them to even higher levels of achievement. The purpose of the latter is to ensure, as much as possible, that the teachers achieve an adequate understanding of standard classroom mathematics. Attention will therefore be focused on bread-and-butter topics in school mathematics, though they will be presented from a slightly more advanced point of view." (Wu, 1999).

In this direction, Jay Jahangiri, in collaboration with three other Kent State University faculty, is under a grant to work with both in-service teachers and prospective teachers in two areas of mathematics; namely, Statistics and Geometry. The classroom approach will be *inquiry- based* presentation. This experimental approach will focus on learning-by-discovery method rather than lecture delivery. The final report and findings of this project will be published in due course.

On the other front, our focus is on "damage control." After connecting with business/ industry/community, this "damage control" would involve working with the newly hired and the "experienced" workers in bringing their math skills up to the company's desired level for productivity, which would be based upon the company's job performance requirements. The high school and college graduates who are working in local businesses/

industries/communities need this "damage control" remedy. Many employers expect their employees to have a competency in problem solving and reasoning skills as well as being able to read, write, do essential mathematics, and understand the technology utilized within the business/ industry/ community. Increased job competition and new technology make continuous learning imperative for the executive and professional employee's as well.

Working with businesses/industries/communities is an important aspect of what those of us in Workforce Development do. In this direction, the Faculty and Workforce Development Department at Kent State University Geauga Campus have created a partnership to serve the businesses/industries/communities in its service area. Continuous learning is essential to an employees employment future.

In order to balance education and training opportunities, Workforce Development Departments in Ohio's institutions of higher learning are supported by their University/College and also by the Ohio Board of Regents and its Enterprise Ohio Network. We are provided with a facility, materials, training, and often financial support to help companies continuously improve their workforce resulting in greater retention, and ultimately, greater productivity for the business. Any roadblocks to productivity that impede our progress are met with various means to resolve them. Training enhancements will make businesses/industries/communities more competitive with states which already capitalize on the non-credit job training offered by institutions to support new business attraction and expansion to a locality, and remain globally competitive in today's marketplace.

Employers require institutions of higher learning to provide them with quick response, customized, non-credit job training at an affordable cost. These institutions must support local business/industry/community growth by offering targeted, affordable, and applicable training. Growing and changing workplaces require that we use all of the workforce development resources available.

Effective training will help the institutions' service area attract, develop, and retain — through training — companies strategically important to the state's economy. The results will be better job retention and the expansion of a state or municipality as a business location for targeted industries; greater use, and impact, of the campuses as a far-reaching economic development resource, or just for better business/industry/community overall performance. Based on interviews with employers, it is estimated that as many as 3 out of 4 jobs require some post-secondary education.

The challenges to companies are great. Companies that are hesitant to train employees because they feel that their newly trained employee will leave that company to go to a competitor, often do not realize what happens if they *don't* train and develop their employees. This can be as devastating to a company as NOT doing *any* training at all. Those already in the labor force must adapt to the technology needs of a highly competitive global economy. Workers are working longer and retiring later. As a result, older workers must continue to polish skills and gain new skills in order to stay informed and important to the company.

Companies are providing employee benefits to attract and retain workers, and continuing education and developmental education are drivers. This provides employers the assurance that their employees are prepared for higher levels of job responsibilities and expectations.

There are three ways to accomplish eliminating the "Alien" concept that many institutions of higher learning seem to unknowingly foster.

1. Encourage companies to institute a Tuition Assistance Program for *all* employees.

2. Institute specialized training incentives or initiatives within the walls of the company which would encourage employees who don't have a high school diploma to investigate and take part in developmental skills which would lead to that employee's earning a GED. With new education levels, that employee could advance to college-level work and be able to take advantage of the Tuition Assistance Program.

3. It is very important that the faculty become "user friendly" to the *entire* learning audience. Effective programs are not limited to developmental skill areas.

A trained workforce will encourage more businesses to incorporate Tuition Assistance Programs in the benefit plans for their employees. With more Tuition Assistance Programs available to employees, resulting in a better-trained workforce and campuses will see an *increase in enrollment* in credit classes and degree programs. It's a WIN/WIN situation for EVERYONE! We must continue the momentum... We must provide training that is effective and affordable, and then be accountable for the outcome. Continuous learning through *continued learning* should be the objective of all workforce development departments, which could utilize the expertise of the faculty members. We must also be aware that many employees who participate in continuous learning may already have a degree, but may still need developmental programs!

Once a company implements effective training procedures, and workers experience the value it has added to their life—both work and personal—the institution becomes "friendly"—no longer an *alien*. In this instance, "friendly" is the objective to rejecting the *myth* that universities and colleges are *foreign/intimidating* to the people and community they are supposed to serve.

We recently worked with several local manufacturing companies wanting "assessment testing" of their employees. This came about as a result of the QS-9000 certification process that the companies were *FORCED* to obtain. (QS-9000 is a series of compliance standards a company must meet. For example...when you see the UL logo on an electrical appliance, you know that that particular item is SAFE because it had to meet certain standards established by the Underwriter's Laboratory. Same with QS.) These companies, while pursuing the QS-9000 certification, found that during the QS mandatory training, employees were unable to follow basic instructions, and had difficulty understanding what the instructor was teaching because the trainees didn't have the math skills needed to perform certain functions required in the QS training. The employees also had difficulty communicating via the written word—reading instructions, writing procedures, etc.

Using a generic one-hour "locator" test consisting of Reading, Language, Spelling, Math Computation and Applied Math — followed with a 2.5 hour "survey test" covering the same topics—it was determined which "Survey" test the employees would later take...which would then determine which developmental classes should be provided for the employees at the company. (The testing instrument that was selected for this particular company consisted of two parts: Locator Test—which was used as the "Pre-test" — and a Survey Test. These "generic" testing instruments are available through educational publishing companies. All Workforce Development and Continuing Education Departments have access to these instruments.)

Many people were afraid of the test and some were quite vocal... "Why do we have to do this?" "I know I'm going to lose my job..." "I'm almost ready to retire."

EVERYONE from the Plant Manager to the newest shop employee participated in the testing process.

The resulting findings of the Survey test were interesting. Some people had difficulty with reading, many couldn't do the basic math skills, and many couldn't do the applied math.

Throughout the testing the employees were assured that the test results would remain confidential and stay with the University. Management was cooperative and understanding, and did NOT even ask to see the scores. After the Survey test, we met with the employees who had taken the test to give them their results. Three of the most frequently missed math problems were explained. The pre-testing hostility soon turned into a "friendly" atmosphere with the employees asking questions and expressing "surprise" at how the answers were derived. We made the math interesting and relevant to the learners by using basic, easy-to-understand examples.

Our positive experience in teaching these classes proved that meaningful mathematics learning results only if the students are *actively* involved in personal sense-making by using pertinent examples as mentioned below. We encouraged students to form groups of three or four. In very rare occasions, groups of two were allowed depending upon the students' subject matter background. The groups were mainly heterogeneous, but we were not "picky" in regard to allowing the students to form a group with which they felt comfortable while working together. To this structural formation, we applied the *inquiry-based method* for both classroom instruction and assessment of student learning. By "inquiry-based method" we mean that in both instruction and assessment of students' learning, in particular, we considered the general mental mechanism for thinking and learning, including, but not limited to, such things as abstraction, reflection, and socio-cultural factors.

For example, using the strategy of real-life experiences and applications such as... "I am pedaling a bicycle from Mentor (the location of their plant), and I start out 4 miles from Mentor...." was used to explain a math problem using negative numbers. IT WORKED! Interest was high, and the same comment came from each group: *NOW I get it!!!* and, *I wish I had a math teacher who had made math interesting when I was in school!*

In this review process, we emphasized conceptual understanding, and exploitation of the connections between mathematics and other quantitative disciplines. The employees were given the opportunity to examine the realistic applications of mathematics and also were challenged with open-ended problem-solving questions. The main, and most important and successful technique, was collaborative learning and the use of technology. We used inquiry-based methods to help these students to discover mathematics and its applications on their own rather than to be fed to them.

As a result of the assessment, developmental math and language classes were offered at the company site. A post-test (the contents of which being the same as the initial *pre-test*) was administered at the end of each session and compared with the pre-test. As a result of the training, in the post-test, almost everybody scored 90% and above, whereas in the pre-test, many scored less than 30%. It must be noted that our agenda was not to teach "to the test" and only instruct on what was covered in the test, but to provide instruction which gave the students the tools to answer and calculate the questions using their new-found mathematics ability.

At the end of each sessions, students were asked to evaluate the course they had just taken. The evaluations were outstanding. One student wrote, *...if I had had a math teacher like you in my grade school years, I would now be working at a higher level job. You make math interesting and meaningful, and something that I can use everyday.*

Interestingly, one of the most vociferous people during the testing was one of the first to sign up and is now working on his GED. He even took time to write a note saying how pleased he was with the class and the instructor. Earning a GED may be just the first step for many of these learners to going on to bigger and better educational opportunities. Several of the people taking the classes have expressed an interest in taking credit courses and working toward an Associate Degree. (It doesn't have to stop with the Associate Degree either.)

For a community project, we had a *College for Seniors* (senior citizens) during the summer. The participants could chose to attend an Internet, math, or a humanities class. EVERYONE came out of the math presentation expressing the same thing: *I wish my math teacher had taught math like that! It was SO relevant.*

Youngsters in the summer *College for Kids* and *College for Teens* programs also had an opportunity to take math classes on campus. The parents were delighted with the progress made.

Another program for youngsters was the *Job Training Partnership Act* Summer Youth program, which was a program for students with a learning disability, some type of physical disability, or belonged to a lower income level. The group discussion and class participation proved successful in JTPA classes. We received many complimentary remarks. For example, one youngster was delighted to learn how to measure correctly! *I used to just guess! Now I know how to really do it!* His lumber yard employer was even more delighted than the student was!

The School-To-Work program has been instrumental in helping youngsters, parents, teachers and the workforce communicate. Things will not get better until we communicate the how's and why's of what we are learning in school, how we are teaching our youngsters, and what expectations the workforce places on its employees.

There will always be a place on campuses for faculty/workforce development and continuing education partnerships because there will always be a need for continuous learning. We can't just STOP! Learning is truly a *Cradle-to-Grave* process!

As institutions of higher learning, we must make ourselves more palatable...more friendly...not a "force to be reckoned with." We must let people know that we are available to help them be successful in whatever endeavor they choose while maintaining the dignity and respect associated with education and the institutions providing it. We must be all things to *all* people in one way or another. We must be a provider of continuous learning for everyone...a haven for those in need of greater learning, no matter what the need of the potential student. We must be a resource...a tool...to keep our great system functional and in touch with reality. [There is a greater number of immigrants entering this country and joining the workforce. We must be ready to "fill in the gaps" they may have in mathematics.] Our businesses/industries/communities support our campuses with scholarship dollars...this must be our way of "giving back" and showing that we appreciate these efforts.

Bibliography

H. Wu, Professional Development of Mathematics Teachers, Notices of AMS, May 1999, 535 -542.

10

SUPPORT SYSTEMS IN BEGINNING CALCULUS*

Myrtle Lewin
Thomas W. Rishel

Background

Cornell University's calculus sequence for non-majors in the School of Arts and Sciences is a traditional course, with several sections of under 25 students meeting four times a week, and taught by Cornell faculty, visitors and graduate student teaching assistants. Academic support is available from several sources: a Mathematics Support Center, offering peer tutoring in the Department; a Learning Skills Center, which was originally formed to support students from underrepresented groups, but which offers assistance and evening peer tutoring in a variety of disciplines to all who come; student graders; and instructors' office hours.

Lewin, a 1992/1993 sabbatical visitor at Cornell from a small women's liberal arts college, was scheduled to teach two sections of Calculus I in the spring semester. Rishel was the coordinator of the Calculus I course that semester. The coordinator, affectionately known as "The Czar" of the course, makes up the syllabus and homeworks, puts together exams and makeups, and generally acts as "ultimate arbiter" in disputes. Lewin wanted to take advantage of the relatively large numbers of students and gender mix to implement some of the techniques of collaborative learning that she has practiced at her home institution, in the hope that some gender data would be collected.

We recognized the contrast between the strong mentoring atmosphere of the instructor-student relationship in a small college, and the fairly extensive alternate support system that exists at Cornell, where the only informal contact between instructor and student is during instructor's office hours. These hours are relatively poorly utilized by Cornell students who tend to be extremely respectful of their professors, and who therefore experience very little unstructured contact with instructors in beginning calculus.

Together, we planned to introduce twice weekly voluntary two-hour study sessions staffed by a team of course instructors. In these sessions, we would encourage our students both to use each other as resources, and interact informally with instructors. Sessions were held on Wednesday and Sunday evenings, two hours each, in a large classroom with moveable chairs. Students were free to work independently, or to gather in small clusters, while instructors move among them responding to their requests for help. Classroom chalkboards were not used, so that the sessions would take on the aspect of discussion rather than lecture. We would then observe student study patterns. In every other respect, the course would run along traditional lines: instructors would have autonomy in the classroom, would give their own quizzes and hold office hours. The three tests (prelims) and final exam were timed, closed book events, composed by the coordinator and graded by all instructors, and final grades were to be determined by consensus among the instructors, but on a curve with an average of C+/B-. Participation by instructors in these study sessions was voluntary; all but one attended at least a few of the sessions. A schedule for instructors was arranged in anticipation of needs, and in order to ensure that at least two instructors would be present at each session. The two authors each attended the vast majority of the sessions, often together, and there were varying levels of participation by other instructors. But each was encouraged to attend at least three or four of the sessions "to find out what your students are having trouble with." We discussed with instructors the role that these study sessions could play not only in their students' learning, but in their own development as teachers, and asked that all instructors encourage their students to attend these sessions. We also provided instructors with rather firm guidelines for their participation in these study sessions: no blackboard use; the students should always have paper and pencil in hand; we were to listen carefully to a student before intruding with our own ideas; we were not to over-help; we were to encourage coherent writing; and we were not to rob our students of the excitement or frustration of discovery and false starts.

The Study Section

These study sessions developed a dynamic that was interesting. Attendance varied, but there were several students who attended regularly, some even twice weekly. The level of participation also varied, from students who would pop in for a few minutes, knowing that they had specific problems to deal with, to those who would come in for the full two hours and seldom interact with anyone. Each session was staffed by two or three instructors, in a large classroom. We asked students to sign in.

Instructors moved around in response to requests from students, sitting down next to a student or group of students engaged in discourse, asking questions, giving a hint or making a suggestion, and then moving on, frequently returning to follow up on the progress of a problem or the development of understanding.

We instructors sometimes chatted with each other, discussed approaches we had taken or planned to use in our classes, and shared ideas. We learned to work with each other's students, and our students worked with friends in different sections.

One of our objectives in establishing these sessions, unwritten and unspoken except to each other, was to encourage instructors to observe their students engaged in study so that they would be motivated to adopt alternate approaches in the classroom. When the teacher never sees students struggling, s/he has no reason to doubt the efficacy of the lecture method, no reason to acknowledge that different students, through no fault of their own, might be exposed to very different educational experiences in the classroom. Further, where contact between instructor and student is limited to fifty-minute capsules or brief "how do you do this problem?" questions during office hours, there is little incentive for dialogue.

By being forced to listen carefully to students' half-formed ideas and to read their jottings, rather than showing them how clever we were, we often learned where students were confused. We regularly integrated conversations we had in these study sessions with the content of our later classes, making it clear to all our students that we were learning from their efforts, that there are often multiple approaches to a problem and that we as instructors valued the insights that our students offered. This was valuable psychological reinforcement for both instructor and student. As course coordinator, Rishel was often able to point out to the instructors items that students were still having trouble with, and to suggest strategies for overcoming these problems in class. We improved our capacity to allow students time to "reflect or to collect their thoughts before answering questions addressed to them" ([3, p. 151]). We could also prevent a class from being derailed by suggesting that we chat about a topic or question of interest to only one or two students in the next study session. Particularly for those instructors who were regular study session participants, these study sessions became an extension of the classroom — a wonderful side benefit!

The presence of at least two or three instructors at each study session had an added bonus. Students had the opportunity to show their inadequacies to, and to be mentored by, an instructor who was not also the judge and assessor of their work. In this way we were able to separate the teaching function from the evaluative function of an instructor.

A more subtle objective was to deal with those gender stereotypes which may have diminished in our society but have not disappeared. It is well documented that many young women still believe that to participate and succeed in mathematics they will have to pay a price. The price is higher in the high school years when prevailing stereotypes are felt more strongly. In [5], Livson and Peskin suggested that so-called "sex-inappropriate behavior [among young women] in early adolescence is an expression of protest against the incompetency 'demands' of the conventional feminine role."

More important, instructors may unknowingly be vulnerable to gender discriminatory behavior. Myra and David Sadker point out that, even in college and graduate school, significantly more attention is given to men than to women, and spontaneous (assertive?) behavior is more tolerated from men than from women students [7]. Furthermore, as Mary Schatz Koehler points out ([3, p. 145]), "teachers may have been enabling males to become independent, while unknowingly not enabling females to achieve autonomy."

We took advantage of the need to provide instructors with guidelines describing the atmosphere we hoped to create in these study sessions by discussing with them some aspects of current research on gender-differentiated behavior. We believe that by increasing their (and our) awareness of these issues, and by practicing what we preached, we discouraged behavior by instructors that may have had a gender bias. Further, participating in these study sessions were students and instructors of both genders. The classroom atmosphere, recognized as being so vulnerable to gender-discriminatory behavior, had been replaced by an alternate learning environment which fostered both collaboration and independence.

Our Survey

We gathered information from each student in two narrative surveys, collected by instructors in each section at the beginning and end of the semester. The first, in the first week of classes, asked for some demographic data, mathematical history, anticipated course grade, and long term plans. The second, in the last week, asked for long term plans and anticipated course grade (were these different?), and their perceptions of, and reactions to, the various support mechanisms available, including the study sessions. We also asked whether students had participated in any informal student initiated study groups (in the dorms, fraternity and sorority houses, or other contexts). These are referred to as *private working groups* in what follows.

We compared narrative responses on these two surveys for each student, hoping to:

a) Gauge any changes in attitude to mathematics and its usefulness to the student during the semester;

b) Correlate data on any prior calculus experience with performance and perceived success, and do this by gender;

c) Compare participation in, and perceived usefulness of, support systems (including the study sessions, and private working groups) with prior calculus experience and gender.

Our Findings

We distinguished three groups of students by prior calculus experience:

1. No prior calculus;
2. A non-AP calculus course;
3. An AP course, either AB or BC, whether the exam was written or not;

and refer to these three groups as Groups 1, 2, and 3 below. A student with a 3 in the AB, or a 2 in the BC, would have placed out of Calculus I at Cornell. We also use the notation W = # of women, M = # of men.

The vast majority of students who dropped the course were in Group I (30 of the 43 who dropped), but these students were distributed between the two genders (M = 24, W = 17, 2 genders unknowns) in roughly the same proportions as among those who completed the course (M = 58, W = 45).

The comments that follow apply to those students who completed the course, and for whom a final grade was reported. Further, because the surveys were narrative, allowing students to leave blanks or respond ambivalently, we have different numbers of responses to different questions. This accounts for what sometimes appears to be inconsistent reporting.

First, students with no prior calculus (Group 1) performed as well, and felt as good about the experience, as did those with a prior AP course (Group 3). But students in the 'no prior calculus' group used all the support mechanisms more, and worked harder, than did those in the AP group. By contrast, students with a prior non-AP calculus course (Group 2) were least inclined to use any of the facilities provided, had lower expectations for course grade, and did not perform as well, as did students in either of the other two groups.

We gathered from the entry survey that, while more than half of the respondents with no prior calculus (13 out of 23) felt that they were inadequately prepared for the course (only one out of 18 respondents with AP calculus felt this way), very few of them anticipated a grade of less than B (5 out of 36 in Group 1, compared to 8 out of 45 in Group 2 and 1 out of 28 in Group 3).

When we looked at the relationship between gender, prior calculus experience, and grade distribution, we found some very interesting differences.

On the entry survey, 9 out of 45 women predicted a grade of less than B, compared to only 3 out of 58 men. And yet the women fared as well as, and in fact better than, the men. (Average grade on a 4 point scale: 2.72 for women, 2.57 for men). This is accounted for in part by the fact that far more men than women received F's (M=7 compared with W=2). One possible explanation for this is the social prejudice discussed earlier. It would appear that a woman student, anticipating failure, would be more inclined to drop a course rather than fail it, whereas a man student would be more willing to take a chance, and fail. This is consistent with the well-documented view that, particularly in mathematics, men tend to attribute failure to lack of effort, whereas equally bright women tend to attribute failure to lack of ability ([2, pp. 1044 – 1045]).

Possibly the most interesting finding was the breakdown into the three groups of prior calculus experience by gender:

Group 1	38 students	F=12	M=26
Group 2	46 students	F=21	M=25
Group 3	28 students	F=16	M=12
Totals	112 students	F=49	M=63

It appears that prior calculus experience and gender are not independent. Certainly, women students at Cornell appear to be more susceptible to the belief that calculus is a prerequisite for calculus. The high proportion of women with prior AP calculus may account for the higher final average grade we reported. This raises the question as to whether these women then continued into Calculus II in the same proportions.

Particularly at a school of the caliber of Cornell, where AP credits are abundant, there may be the perception on the part of women who have not been accelerated in mathematics in high school that they should be cautious about their mathematics course choices in college. This perception may be built into the advising system, or it may be self-advising. But it should make us all take note, and not perpetuate the perception that women need not advance their mathematical skills with the same sense of urgency as men.

Apropos this last, Rishel noted a significant indicator at drop-add day ("Grand Course Exchange") last semester. Four women, acting separately over a period of four hours, signed into Calculus I, only to return later to cancel their registration. Each, when asked by the administrator (who was a woman) why she wanted to do so (no men had done this), said she had been told by other students that the course would be "too hard." Even after prodding, none of the four students would change her mind about not even attempting the course.

Lynn Arthur Steen [8] reminds us that "The learning of mathematics entails profound socio-political consequences. Success or failure in mathematics determines access to courses and curricula that lead to positions of influence in society."

Support Mechanisms

Our examination of participation in support mechanisms was equally interesting. Participation was defined loosely as having used the mechanism more than once, and having reported some benefit.

It appears that students with no prior calculus made better use of the study sessions and peer tutoring than did those in the other two groups, but were least inclined to participate in private working groups. Essentially all students who completed the course turned in surveys, so we report these results as percentages. We found that 55% of students in Group 1 used the study sessions, compared to 21% of those in Group 2 and 35% of those in Group 3.

We expected students with prior AP calculus (Group 3) to be self-sufficient and competitive. But surprisingly, 85% of these students reported using private working groups, compared to 54% in Group 2 and 52% in Group 1.

We found that women used all forms of collaborative support more than did men. Among those who returned the second survey, 44% of women reported using the study sessions, compared to 35% of the men. And 69% of women reported using private working groups, compared to 54% of the men.

These latter numbers are interesting, because in [4], an assessment of undergraduate learning at Harvard in 1990, it was suggested that "the many men and few women who form study groups report that they both enjoy their work more, and feel they learn more, because of the academic discussions within these groups...Ironically,...women...are far less likely than men to join one or start one."

We have no information on the gender mix in these private working groups. But we would like to believe that our encouragement of our students to study with others in the same course (rather than being helped by friends in more advanced courses) had something to do with the extensive use of private working groups. Whether this is true is open to speculation and, we believe, warrants further study.

Short Term Effects

Rishel is again Coordinator of the Calculus I course this school year. Based largely on this study, he has initiated a number of changes in the structure of the course.

Study sessions continue. Further, much more attention is now being paid to the majors and career goals which students announce in their entry surveys. Consequently, business and biological applications are more prominently featured, because about 40% of the students expect to go into business or economics, and almost one third say they "are considering" medical school.

At the same time, instructors stress the usefulness of "continuing a little further into mathematics." An example of such a situation is as follows: A simple Calculus I example is done, after which the instructor says "This problem may not be so hard, but here is a 'real world' situation just like it... This real world problem can be written in Calculus II language, and can be solved using Calculus III techniques, when you take that course." For instance, a discussion of solving

$$dy/dt = ky$$

during a lesson on logarithms leads to a mention of the variables-separable technique of solving differential equations. This, in turn, allows the instructor to show the logistic growth equation and discuss what that equation has to do with ecological systems. No derivation need to be done here ("You'll see the details in later courses") to motivate the concept that mathematics and biology are intertwined.

This year, for the first time, more women than men are taking the Calculus I course. Rishel has found that last year's and this year's interactions with women instructors and students has let him to a deeper understanding of who his audience is; which responses are enabling and which aren't, and how this interaction does or does not motivate students, in particular women, to continue study in the field of mathematics. He also now has a better idea of what students might really be asking when they want to see, say, an implicit differentiation problem. By their tone of voice, by the help they give Rishel as they go through the solution together, the students can show whether their question is purely one of mechanics, or whether they still haven't absorbed the chain rule.

Conclusion

This study is preliminary and incomplete, and the mechanisms for data collection and analysis were not refined. But we feel that we uncovered enough of interest to report to an audience beyond our departments. We believe that studies of this type, which can lead to changes in policy of student advising or course structure, are worthwhile. And we learned about our own teaching because we were simultaneously playing the role of observer and participant in the study.

As we read more and more about the "subtle changes in emphasis [which] are giving way to substantive and substantial changes in defining what calculus reform is truly about, ... ([T]eaching and learning are the new hot issues, emphasizing student centered environments." [6]), we would offer some advice to those who want to introduce alternative instructional approaches, but who are constrained by syllabus, departmental structure, or demands of uniformity.

• By introducing study sessions, one creates an environment in which the students become involved actively in the learning process, and in which the atmosphere in the classroom reflects this involvement. But one does not have to compromise significantly on content. One really has the best of both worlds.

• In addition, the environment of the study sessions, guided by the principles of collaborative learning, provides an opportunity for mentoring of less experienced instructors to take place in a non-threatening context.

As long as instructors use their presence wisely, pulling others into their conversations, showing the value of careful listening, of respect of the student's need to flounder, and of effective nudging, and practicing habits of not over-helping, study sessions will be a valuable tool for students and instructors to learn together.

* Reprinted with permission from PRIMUS-Problems, Resources, and Issues in Mathematics Undergraduate Studies. Vol. V, No. 3, pp. 275-285, Sept. 1995. Published at the United States Military Academy, West Point NY.

References

Belenky, Mary, et al. 1986. *Women's Ways of Knowing.* New York: Basic Books.

Dweck, Carol S. 1987. Motivational Processes Affecting Learning. *American Psychologist* 41: 1041-1048.

Fennema, Elizabeth and Gilah Leder, ed. 1990. *Mathematics and Gender.* New York: Teachers College Press.

Light, Richard J. 1990. *Harvard Assessment Seminars: Explorations with Students and Faculty about Teaching, Learning and Student Life - First Report.* Cambridge MA: Graduate School of Education and Kennedy School of Government, Harvard University.

Livson, N. and H. Peskin. 1981. Psychological health at age 40; predictions from adolescent. Eds D. Eichorn, J.A. Clausen, N. Hahn, M.P. Honzik and P.H. Mussen. New York: Acdemic Press.

Mathematicians and Education Reform Network, Newsletter. 6(1), Fall 1993.

Sadker, Myra and David Sadker. 1986. Sexism in the classroom: from Grade School to Graduate School *The Phi Delta Kappan.* March: 11-14. Reprinted in *The Association of Women in Mathematics Newsletter.* 20(6). Nov-Dec. 1990.

Steen, Lynn Arthur. 1992. 20 Questions that Deans should ask their Mathematics Departments. *American Association of Higher Education Bulletin.* May 1992. 3-6.

11

A COMPONENT SYSTEM FOR COLLEGE MATHEMATICS AND STUDENT READINESS

Ron C. Goolsby
Thomas W. Polaski

Introduction

In this chapter, we discuss a scheduling process which we have used at Winthrop University for the past five years in an introductory level course. The system "splits the difference" between a standard semester-long course and a self-paced skills course by using three components which students must pass through to pass the entire course. As described later, this system differentiates students by skill level as the course proceeds. As a result, we have been able to analyze the differences between students who succeed in the course, students who struggle with portions of the course, and students who fail the course. The results of that analysis and suggestions for improving student readiness for college mathematics are also included in this report. The component system at Winthrop is not unique; other institutions such as the University of Texas at El Paso [1] also use component-type systems which they have developed independently.

The Course

Winthrop University is a comprehensive public university in the state of South Carolina with an entering freshman class of approximately 900 students. Each undergraduate must pass one mathematics course to graduate. Precalculus Mathematics (MATH 101) is the most popular option for this course; in a typical fall semester, about 400 students enroll in MATH 101. We decided to experiment on this course because of its popularity and its clearly defined syllabus, and because of several problems the department and university had with its implementation. These problems included:

• large withdrawal and failure rates in the course;
• semester-long time lags between attempts at the course;
• variability of testing and grading procedures among the faculty.

Tables 1 and 2 place numbers on the concerns which we needed to address. In Table 1, student success rates are shown for the last 8 semesters before we instituted the new system. Failure and withdrawal rates for MATH 101 range from 35.7% to 50.2% of the student pool. Table 2 shows the variability in student success rates across sections of roughly the same size taught in the same semester. Rates vary widely, from 35.7% to 69.3%.

Table 1: Enrollments and Success Rates, Math 101
Fall 1990 Through Spring 1994

Semester	Enroll-ment	% Students with A, B, or C	% Students Passing	% Students Withdrawing or Failing
Fall 90	405	41	58.3	41.7
Spr. 91	177	33.9	51.4	48.6
Fall 91	407	36.9	51.4	48.6
Spr. 92	207	29.5	49.8	50.2
Fall 92	431	50.8	64.3	35.7
Spr. 93	205	33.7	52.2	47.8
Fall 93	470	38.8	51	49
Spr. 94	181	35.9	50	50
Total	2483	39.2	54.5	45.4

**Table 2: Success Rate (% A, B, C, or D) for
Individual Sections, Fall 1993
Class Size 37-44 Students**

62.7%	68.8%	51.2%	66.6%	52.5%	39.5%	35.7%
57.7%	51.2%	52.8%	69.3%	37.8%	40.5%	41.2%

To address these concerns, we worked to change the format of the course to what we call a component system.

How the Component System Works

The syllabus, and thus the semester, is divided into three components of nearly equal length. We label the components of the syllabus Components A, B, and C. Each component is designed to be either 13 or 14 class sessions long, and all sections of the course are scheduled for Monday, Wednesday and Friday sessions. Each component is designed to end on a Wednesday; the students are given the following Friday off. The content of each component stays the same even if the number of class sessions varies; review days are added or subtracted as necessary to make everything add up correctly. Examples of the schedules for the Fall 1999 semester are given in Appendix A.

Most students beginning the course in a given semester begin in Component A. At the end of the first third of the semester, all of these students are given a common final examination covering Component A. A numerical grade for Component A is computed using this examination and instructor's input. The final counts for 75% of the grade, with the instructor using quizzes, homework, and/or his or her own tests to determine the other 25%. For ease in grading, the final examinations are multiple-choice format. If a student passes the component at a prescribed level, he or she begins Component B; if a student fails the component, the student must begin Component A again immediately. Students who pass the component at a low level are given both options.

At this point in the semester, the student populations of each section must be shifted to place Component B students with other Component B students and Component A retakers with other Component A retakers. For this reason, all sections of MATH 101 are scheduled at two common times.

Instructors are also shifted; some will teach Component B, while others will reteach Component A. Those instructors teaching Component A for the second time concentrate their instruction on those topics which have given students most difficulty. The department can also control class sizes during this shifting process, often reducing the class sizes of the remedial sections. The process of grading and rescheduling is done internally by the department and takes two days to accomplish, so the students are given one class day (Friday) off while this is done.

Component A retakers now have no hope of completing the course in one semester. These students (provided that they remain engaged in the course) will receive a grade of U in the course. This grade is not used in computing a student's GPA and does not affect the number of earned hours they have attempted. Thus a struggling student's ability to return to the University for a second semester does not depend on his or her performance in MATH 101. They will register for the course again in the following semester and (we hope) complete it then. The student thus has six attempts to pass the three components of the course, and the department has a chance to locate students with little hope of passing the course. During their second semester in the course, these students are advised to remediate their mathematical background before attempting the course again.

At the close of the second third of the semester, the same process occurs again. Now students may end up in Component C, or be guided to retake B, or advance to B after passing A on their second attempt, or take A for the third time. Notice that students are being differentiated by the system into discrete levels of ability; the department's scheduling flexibility can again be used to tailor sections to individual needs.

When the semester is over, those students who are in Component C are given a final grade by averaging their three component grades. Thus a student need not pass Component C to pass the course. Students are given the option of repeating this final component to improve their final grade; they must enroll for another semester to do so. Those students who have not finished the course are placed into the appropriate components for the beginning of the next semester.

Although most students begin MATH 101 in Component A, a few will actually begin the course in Components B or C due to their performance on a departmental placement test administered during Orientation. These students progress through the system as the other students do, and may indeed finish the semester's work in one or two thirds of the semester. This has quite accidentally functioned as an encouragement to the students to perform well on the placement test!

The University's withdrawal system must also be incorporated into the system. This is relatively simple at Winthrop since the end of the first segment of the semester usually coincides with the last date on which a student can drop a course without penalty. In any case, the department guarantees that every MATH 101 student has the opportunity to check his or her performance in the first component before deciding to use the University's withdrawal policy to drop this course. Although students are allowed to drop at this point, they are strongly encouraged to remain active in the course.

Assessment of The System

After five years of working under the component system, we have found that the system has successfully addressed our goals for student ability grouping, for course standardization, and for student retention. Table 3 shows how the students who started in MATH 101 since the fall semester of 1994 have performed. By "active student" we mean a student who stayed in the component system after the first component exam. Most students who are not active withdraw from the course within the first week of classes.

By comparing Tables 1 and 3, the following facts become apparent:

- The percentage of students with an A, B, or C after one semester has actually dropped from its value under the old system (from 39.2% to 30.7%). However, the true percentage of students who now complete MATH 101 with an A, B, or C is 44.9%. This is an increase from the old system.

Table 3: Enrollments and Success Rates of Active Students
Math 101, Fall 1994 through Spring 1998

	Enrollment			% Active Students With A, B, or C		% Active Students Passing	% Active Students With-drawing or Failing
Semester	New	Active	Return-ing	After 1 Sem.	After 2 Sems	After 1 or 2 Sems.	
F 1994	363	323	0	37.2	47.4	65.9	34.1
S 1995	129	104	110	28.8	45.2	60.6	39.4
F 1995	315	281	28	40.6	54.4	65.5	34.5
S 1996	73	65	101	29.2	35.4	44.6	55.4
F 1996	273	255	18	14.1	38.8	56.9	43.1
S 1997	83	69	138	20.3	27.5	37.7	62.3
F 1997	356	327	11	31.2	45.0	56.6	43.4
S 1998	54	47	103	34.0	40.4	48.9	51.1
Total	1646	147	509	30.7	44.9	59.0	41

• The percentage of students passing the course has also increased from 54.5% to 59.0%. The component system has helped a greater number of students to succeed.

• As a result of students progressing more effectively through MATH 101, the enrollments in the course have gone down while Winthrop's total student body has remained fairly constant. Approximately the same number of freshmen register for MATH 101 under the component system as in earlier years, but many fewer upperclassmen must register for the course to retake it. Thus the department has been able to use its faculty to better effect.

This comparison of MATH 101 performance before and after the implementation of the component system shows that this new system has been of help in moving most students through the course and has helped them to perform better in the course. In a later section we will address the students who are not helped by this system.

Problems With The System

Even with the substantial success of the component system, there are problems with its implementation, which must be addressed. We have found that most problems with the system are bureaucratic in nature. The most difficult aspect of the component system is the need to keep track of where all of the students are. Since MATH 101 is still one three-hour course, all of the student and faculty shifting is done by the Mathematics Department. The department developed a spreadsheet with all of this necessary information and assigned one faculty member to shepherd this spreadsheet each semester. This faculty member was given released time to construct and implement the database.

The preparation of component final exams has been problematic. During the initial year (1995) of implementation, the common exams consisted of free response questions that were graded by faculty. Each faculty member was assigned specific questions to be graded, and the individual faculty member devised their own schemes for the awarding of partial credit. A study of the grades resulting from this style of grading showed that the awarding of partial credit had a negligible effect on the students' overall performance. As a result, we began using multiple-choice finals in the fall of 1996. These finals are prepared by one instructor teaching each component and are okayed by the other instructors of that component. Often faculty have had differing understandings of the level of difficulty expected of the students and of the appropriate emphasis of certain topics. In addition faculty have often prepared the finals with little time for other instructors to comment fully on them. To alleviate these problems, a detailed "skills list" for each component was developed to accompany the list of homework problems. This list is included as Appendix B. This list gave faculty guidance as to the appropriate topics for exam questions. This fall we have gone a step farther by accepting the efforts of one faculty member who has volunteered to produce all exams for this semester with plenty of time for corrections and amendments from the other faculty. The department hopes to use these exams as templates for future examinations.

One might assume that students forced to take two semesters to complete the course would be upset about the cost of tuition. In effect they have paid for two semesters of instruction but only gained one semester of credit. We have not heard complaints of this type from the students. Under the old system, most students who now take two semesters would have either failed or withdrawn from the course in at least one semester. A comparison of the success rates for the old system and for the new system bears this observation out. Thus most repeating students would have paid for more than one semester of MATH 101 anyhow while damaging their grade point averages.

Student Readiness

Even with the success of the component system, still about 40% of the students taking MATH 101 either withdraw from the course or fail the course after two semesters of attempts. This is still a high percentage, and it leads us to wonder if the majority of MATH 101 students are actually prepared to take this course. The component system has the happy by-product of differentiating students by levels of success, so it is relatively easy to track cohorts of students with different paths through the system. We decided to take as a population those incoming freshmen in the fall of 1997 who attempted MATH 101. We did not include those students who withdrew from the course, upperclassmen repeating the course, and transfer students. We divided these students into three cohorts:

1. Students remaining in Component A after the second exam of the fall semester (AAA students)

2. Students either advancing to or remaining in Component B after the second exam of the fall semester (AAB and ABB students)

3. Students advancing to Component C after the second exam of the fall semester (ABC students)

Note that the students' record in MATH 101 after the second component exam was not studied. We were mainly interested in how the high school preparation of the students affected their performance in the early stages of MATH 101. However, it is safe to assume that most of the AAA students eventually failed the course while most of the ABC students eventually completed the course successfully. We took a random sample of approximately 50% of each cohort; the results of our data collection are listed in Tables 4, 5, and 6.

Table 4: Fall 1997 AAA Students
Last High School Math Class Performance

Grade Level	Course Title	A	B	C	D	F	Total
Ninth	Algebra II	0	0	0	1	0	1
Tenth	Algebra II	0	0	1	3	0	4
Eleventh	Algebra II	0	2	13	8	0	23
	Algebra II/Trigonometry	0	0	0	0	0	1
	Algebra III/Trigonometry	0	0	2	0	0	2
	Precalculus	0	0	0	2	0	2
Total 9th -11th		1	2	16	14	0	33
Twelfth	Algebra II	1	1	2	2	1	7
	Algebra II/Trigonometry	0	0	0	1	0	1
	Algebra III/Trigonometry	0	0	3	0	0	3
	Precalculus	0	0	1	0	1	2
	Advanced Math	0	0	1	0	0	1
	Advanced Placement Calculus	0	0	1	0	0	1
Total Twelfth		1	1	8	3	2	15
Total All Grades		2	3	24	17	2	48

Table 5: Fall 1997 AAB and ABB Students
Last High School Math Class Performance

Grade Level	Course Title	A	B	C	F	Total
Eleventh	Algebra II	0	1	1	0	3
	Algebra II/Trigonometry	0	0	1	0	1
	Algebra III/Trigonometry	1	1	0	0	2
	Precalculus	0	1	0	0	1
Total 9th-11th		1	3	2	0	7
Twelfth	Algebra II	1	1	1	0	4
	Algebra II/Trigonometry	0	1	0	0	1

	A	B	C	F	Total
Algebra III/Trigonometry	0	2	2	0	4
Precalculus	1	5	5	0	12
Advanced Math	0	0	1	0	1
Statistics	1	0	0	0	1
Advanced Placement Calculus	0	0	1	0	2
Total 12th	3	9	10	0	25
Total All Grades	4	12	12	0	32

Certain trends immediately become apparent. A much higher percentage of the AAA students took their final high school math class in the eleventh grade or earlier. A higher percentage of the AAA students performed at a C or lower level in their final math class. Conversely, a high percentage of the ABC students took math in the twelfth grade and did relatively well in that class, whatever it was. That many students chose not to take math in the twelfth grade is not surprising, nor is the fact that many students did not go beyond Algebra II. The entrance requirements for Winthrop state that three units of mathematics are required for admission, and they must include Algebra I (or an approved substitute), Geometry, and Algebra II. Even though a fourth year of mathematics is "strongly encouraged," it is not required. Students who do this minimal amount of high school mathematics are just following the path of least resistance to college admission.

Table 6: Fall 1997 ABC Students
Last High School Math Class Performance

Grade Level	Course Title	A	B	C	D	F	Total
Tenth	Algebra II	0	0	1	0	0	1
Eleventh	Algebra II	0	1	1	0	0	2
	Algebra III/Trigonometry	0	3	1	0	0	4
	Precalculus	0	1	2	0	0	3
	Advanced Placement Calculus	0	1	0	0	0	1
Total 9th -11th		0	6	5	0	0	11
Twelfth	Algebra II	1	0	0	0	0	1

	Algebra III/Trigonometry	4	3	4	0	1	12
	Precalculus	6	3	6	2	0	17
	Advanced Math	1	1	0	0	0	2
	Probability	0	0	2	0	0	2
	Discrete Math	0	2	0	0	0	2
	Advanced Placement Calculus	3	1	0	1	0	5
Total Twelfth		15	10	12	3	1	41
Total All Grades		15	16	17	3	1	52

The relationship between student performance and behavior in high school and performance in MATH 101 is studied in Table 7.

From Tables 4 through 7 we note:

- Of the students taking Algebra II as their final math class in the eleventh grade, 75% became AAA students in MATH 101.

- Of the students taking their final math class in the eleventh grade, 65% became AAA students in MATH 101.

- Of the students taking Algebra III or higher in the twelfth grade, 60% became ABC students in MATH 101.

- Of the students taking any mathematics in the twelfth grade, 51% became ABC students.

Table 7: Math 101 Performance and High School Math Performance Compared

Students...	AAA	AAB/ABB	ABC	Total
Stopping in any	28	4	3	35
Math in eleventh grade or before	33	7	11	51
Taking Algebra III or more advanced class in twelfth grade	7	20	40	67

Taking any math in twelfth grade	15	25	41	81
Stopping with Algebra II in any grade	37	9	4	50
Stopping with Algebra III in any grade	5	6	16	27
Receiving an A in final math class	2	4	15	21
Receiving a B in final math class	3	12	16	31
Receiving a C in final math class	24	12	17	53
Receiving a D or F in final math class	19	4	4	27
Stopping in eleventh grade with a grade of C or lower	30	3	5	38

- Of the students taking Algebra II as their final math class in any grade, 74% became AAA students in MATH 101.

- Of the students taking Algebra III as their final math class in any grade, 59% became ABC students in MATH 101.

- Of the students receiving an A or a B in their final math class, 60% became ABC students in MATH 101.

- Of the students receiving a C or below in their final math class, 54% became ABC students in MATH 101.

- Of those students having what might be called "minimal high school mathematics for admission" (a C or below in Algebra II taken as their final math class in the eleventh grade, 81% became AAA students in MATH 101.

Since the syllabus for MATH 101 is very similar to that for Algebra II, we can see that the primary reasons for poor performance in MATH 101 are marginal performance in the similar class in high school and a lengthy time interval between their high school and college mathematics classes.

Perhaps more confounding is the difference between the AAA/ABB students and the ABC students. Table 8 lists SAT and class rank data for each group of MATH 101 students.

Table 8: Fall 1997 MATH 101 Students
SAT and High School Class Rank

		AAA	AAB/ABB	ABC
SAT Score	Median	420	510	495
	Interquartile Range	390-480	410-540	450-530
Class Rank	Median	71%	80%	83%
	Interquartile Range	54%-79%	58%-88%	69%-90%

The AAB/ABB students have a higher median SAT than the ABC students, and there is not much difference in their class rank. The similarities are also borne out in the previous tables. Almost exactly the same percentage of each group took mathematics in the twelfth grade. The major difference we find is in the performance level in that twelfth grade class: 48% of the AAB/ABB students received an A or a B in their twelfth grade math class while 61% of the ABC students did. The difference here seems to be one of application: students who do reasonably well in high school mathematics may be lulled into a false sense of security by the fact that MATH 101 is at a lower level than their final high school class. Many students may then ignore the class until their grade begins to suffer; in the component system, they may be compelled to retake a component as a wake-up call.

Conclusion

We have found that instituting a component system for the college algebra course at Winthrop University has increased student success in the course and also increased student retention. Students begin remediation of difficult material immediately with faculty supervision. Consistency of grading across sections of MATH 101 has been improved by the use of block finals and a syllabus which must be consistent. The component system is more difficult to operate than the traditional system it replaced, but the difficulties are more than overcome by its benefits. In addition, we have been able to use the component system to understand the plight of those students who do not succeed in it (and would not have succeeded in the previous system either). These students tend to have the minimal mathematical experience necessary for college and many have left a long time gap between their high school and college mathematics classes.

Appendix A
Math 101 Syllabus — Fall 1999
Component A — First Session

Class #	Day	Date	Section	Topic
1	W	8/25	P. 2	Integral Exponents
2	F	8/27	P. 3	Rational Exponents and Radicals
3	M	8/30	P. 4	polynomials
4	W	9/01	P. 5	Factoring polynomials
5	F	9/03	P. 6	Rational Expressions
6	M	9/06		Review and/or Test
7	W	9/08	1. 1	Linear Equations
8	F	9/10	1. 2	Applications
9	M	9/13	1. 2	More Applications
10	W	9/15	1. 4	Quadratic Equations
11	F	9/17	1. 5	Linear and Absolute Value Inequalities
12	M	9/20	1. 6	Quadratic and Rational Inequalities
13	W	37155		Review
Final Exam	W	37155		

Math 101 Syllabus
Component B — First Session

Class #	Day	Date	Section	Topic
1	W	8/25	2.1	The Cartesian Coordinate System
2	F	8/27	2.2	Functions
3	M	8/30	2.3	Graphs of Relations and Functions
4	W	9/01	2.4	Transformations and Symmetry of Graphs
5	F	9/03	2.5	Operations with Functions
6	M	9/06		Review and/or Test
7	W	9/08	2.6	Inverse Functions
8	F	9/10	2.7	Variation
9	M	9/13	3.1	Linear Functions
10	W	9/15	3.1	Linear Functions
11	F	9/17	3.2	Quadratic Functions
12	M	9/20	3.2	Quadratic Functions
13	W	37155		Review
Final Exam	W	37155		

Math 101 Syllabus
Component C — First Session

Class #	Day	Date	Section	Topic
1	W	8/25	3.3	Zeros of Polynomial Functions
2	F	8/27	3.5	Miscellaneous Equations
3	M	8/30	3.6	Graphs of Polynomial Functions
4	W	9/01	3.6	Graphs of Polynomial Functions
5	F	9/03	3.7	Graphs of Rational Functions
6	M	9/06	3.7	Graphs of Rational Functions
7	W	9/08		Review and/or Test
8	F	9/10	4.1	Exponential Functions
9	M	9/13	4.2	Logarithmic Functions
10	W	9/15	4.3	Properties of Logarithms
11	F	9/17	4.4	Equations and Applications
12	M	9/20	4.4	More Equations and Applications
13	W	9/22		Review
Final Exam	W	9/22		

Math 101 Syllabus
Component A — Second Session

Class #	Day	Date	Section	Topic
1	M	9/27	P.2	Integral Exponents
2	W	9/29	P.3	Rational Exponents and Radicals
3	F	10/1	P.4	Polynomials
4	M	10/4	P.5	Factoring Polynomials
5	W	10/6	P.6	Rational Expressions
6	F	10/8		Review and/or Test
7	W	10/13	1. 1	Linear Equations
8	F	10/15	1. 2	Applications
9	M	10/18	1. 2	More Applications
10	W	10/20	1. 4	Quadratic Equations
11	F	10/22	1. 5	Linear and Absolute Value Inequalities
12	M	10/25	1. 6	Quadratic and Rational Inequalities
13	W	10/27		Review
Final Exam	W	37190		

Math 101 Syllabus
Component B — Second Session

Class #	Day	Date	Section	Topic
1	M	9/27	2.1	The Cartesian Coordinate System
2	W	9/29	2.2	Functions
3	F	10/1	2.3	Graphs of Relations and Functions
4	M	10/4	2.4	Transformations and Symmetry of Graphs
5	W	10/6	2.5	Operations with Functions
6	F	10/8		Review and/or Test
7	W	10/13	2.6	Inverse Functions
8	F	10/15	2.7	Variation
9	M	10/18	3.1	Linear Functions
10	W	10/20	3.1	Linear Functions
11	F	10/22	3.2	Quadratic Functions
12	M	10/25	3.2	Quadratic Functions
13	W	10/27		Review
Final Exam	W	10/27		

Math 101 Syllabus
Component C — Second Session

Class #	Day	Date	Section	Topic
1	M	9/27	3.3	Zeros of Polynomial Functions
2	W	9/29	3.5	Miscellaneous Equations
3	F	10/1	3.6	Graphs of Polynomial Functions
4	M	10/4	3.6	Graphs of Polynomial Functions
5	W	10/6	3.7	Graphs of Rational Functions
6	F	10/8	3.7	Graphs of Rational Functions
7	W	10/13	2.6	Review and/or Test
8	F	10/15	4.1	Exponential Functions
9	M	10/18	4.2	Logarithmic Functions
10	W	10/20	4.3	Properties of Logarithms
11	F	10/22	4.4	Equations and Applications
12	M	10/25	4.4	More Equations and Applications
13	W	10/27		Review
Final Exam	W	10/27		

Math 101 Syllabus
Component A — Third Session

Class #	Day	Date	Section	Topic
1	M	11/01	P.2	Integral Exponents
2	W	11/03	P.3	Rational Exponents and Radicals
3	F	11/05	P.4	Polynomials
4	M	11/08	P.5	Factoring Polynomials
5	W	11/10	P.6	Rational Expressions
6	F	11/12		Review and/or Test
7	M	11/15	1.1	Linear Equations
8	W	11/17	1.2	Applications
9	F	11/19	1.2	More Applications
10	M	11/22	1.4	Quadratic Equations
11	M	11/29	1.5	Linear and Absolute Value Inequalities
12	W	12/01	1.6	Quadratic and Rational Inequalities
13	F	12/03		Review
14	M	12/06		Review
Final Exam				Date and Time TBA

Math 101 Syllabus
Component B — Third Session

Class #	Day	Date	Section	Topic
1	M	11.01	2.1	The Cartesian Coordinate System
2	W	11/03	2.2	Functions
3	F	11/05	2.3	Graphs of Relations and Functions
4	M	11/08	2.4	Transformations/Symmetry Graphs
5	W	11/10	2.5	Operations with Functions
6	F	11/12		Review and/or Test
7	M	11/15	2.6	Inverse Functions
8	W	11/17	2.7	Variation
9	F	11/19	3.1	Linear Functions
10	M	11/22	3.1	Linear Functions
11	M	11/29	3.2	Quadratic Functions
12	W	12/01	3.2	Quadratic Functions
13	F	12/03		Review
14	M	12/06		Review
Final Exam				Date and Time TBA

Math 101 Syllabus
Component C — Third Session

Class#	Day	Date	Section	Topic
1	M	11/01	3.3	Zeros of Polynomial Functions
2	W	11/03	3.5	Miscellaneous Equations
3	F	11/05	3.6	Graphs of Polynomial Functions
4	M	11/08	3.6	Graphs of Polynomial Functions
5	W	11/10	3.7	Graphs of Rational Functions
6	F	11/12	3.7	Graphs of Rational Functions
7	M	11/15		Review and/or Test
8	W	11/17	4.1	Exponential Functions
9	F	11/19	4.2	Logarithmic Functions
10	M	11/22	4.3	Properties of Logarithms
11	M	11/29	4.4	Equations and Applications
12	W	12/01	4.4	More Equations and Applications
13	F	12/03		Review
14	M	12/06		Review
Final Exam				Date and Time TBA

Math 101 Homework Assignments
Component A

TEXT: College Algebra and Trigonometry by Mark Dugopolski, Second Edition. Reading: Addison-Wesley, 1999.

Section	Problems
P.2	1, 3, 5, 7, 9, 19, 21, 25, 27, 29, 31, 37, 39, 41, 43, 45, 47, 49, 51, 53, 55, 57
P.3	1, 5, 9, 11, 15, 19, 23, 25, 27, 29, 31, 33, 35, 39, 41, 43, 45, 47, 49, 51, 53, 55, 57, 59, 61, 63, 65, 67, 69, 71, 73, 75, 77, 79, 81, 83, 85, 87, 89, 91
P.4	7, 9, 13, 15, 17, 21, 27, 35, 37, 41, 47, 55, 57, 61, 65, 69, 73, 77, 81, 115, 117
P.5	9, 11, 13, 15, 17, 19, 21, 23, 25, 27, 29, 31, 33, 35, 37, 39, 41, 43, 45, 47, 49, 51, 53, 57, 59, 61, 63, 65, 67, 69, 71, 75, 77, 79, 81, 85, 87, 89, 91, 93
P.6	1, 3, 5, 7, 9, 11, 13, 15, 17, 19, 21, 23, 25, 29, 31, 33, 35, 37, 39, 41, 43, 45, 47, 49, 51, 53, 55, 57, 59, 65, 67, 69, 71, 89, 90, 91
1.1	1, 4, 7, 11, 15, 17, 19, 25, 27, 29, 31, 33, 35, 37, 39, 41, 57, 59, 61, 67, 69, 71, 73, 75, 77, 79, 81, 86
1.2	1, 3, 5, 7, 13, 15, 17, 19, 21, 29, 35, 37, 39, 41, 46, 47, 49, 63, 69, 70, 73, 75, 77, 79
1.3	1, 3, 5, 7, 9, 11, 13, 15, 17, 19, 21, 23, 25, 26, 27, 29, 31, 32, 33, 35, 37, 39, 41, 43, 45, 47, 49, 51, 53, 59, 61, 63, 65, 71, 73, 75, 83, 87, 93, 95, 97
1.4	1, 3, 5, 7, 11, 13, 15, 18, 25, 29, 32, 41, 47, 49, 52, 55, 61, 62, 63, 65, 69, 71, 89, 91, 93, 98
1.5	9, 11, 15, 20, 29, 32, 35, 39, 39, 52, 55, 44, 62, 65, 85, 86

Math 101 Homework Assignments
Component B

TEXT: College Algebra and Trigonometry by Mark Dugopolski, Second Edition. Reading: Addison-Wesley, 1999.

Section	Problems
2.1	11, 15, 17, 23, 25, 33, 39, 41, 53, 54, 57, 59, 65, 67, 69, 75, 77, 79, 81
2.2	1, 5, 7, 15, 35, 37, 39, 41, 48, 49, 51, 53, 63, 72, 77, 79, 81, 83, 87, 91, 94, 95, 105
2.3	3, 6, 9, 11 15, 18, 23, 24, 26, 31, 32, 33, 35, 37, 41, 43, 47, 49, 52, 53, 54, 55, 56, 57, 61, 63, 69, 70, 71, 72, 73, 74, 84, 90, 91
2.4	1, 2, 3, 5, 7, 9, 11, 13, 15, 16, 17, 18, 19, 20, 21, 22, 23, 25, 27, 29, 31, 35, 41, 43, 47, 49, 51, 57, 63, 65, 66, 67, 68, 69, 70, 71, 72, 73, 75
2.5	1, 3, 5, 7, 13, 15, 19, 33, 39, 40, 41, 43, 49, 51, 53, 55, 65, 67, 83, 85, 87, 89, 99,
2.6	3, 7, 9, 13, 17, 18, 19, 20, 21, 22, 43, 45, 49, 53, 55, 57, 59, 61, 63, 83, 84, 85, 86, 87, 93
2.7	1, 3, 5, 7, 9, 11, 13, 15, 17, 19, 21, 23, 25, 27, 29, 31, 33, 35, 49, 51, 53, 55, 57, 59, 61, 63, 65
3.1	1, 4, 7, 11, 15, 16 19, 25, 27, 28, 29, 31, 35, 38, 41, 45, 50, 51, 53, 54, 55, 57, 59, 62, 64, 65, 81, 83
3.2	1, 3, 5, 14, 19, 20, 21, 25, 27, 28, 33, 43, 45, 46, 47, 48, 49, 50, 53, 53, 54, 63, 65, 66, 83, 85

Math 101 Homework Assignments
Component C

TEXT: College Algebra and Trigonometry by Mark Dugopolski, Second Edition. Reading: Addison-Wesley, 1999.

Section	Problems
3.3	1, 3, 5, 31, 33, 35, 37, 39, 41, 43, 45, 47, 49, 51, 53, 55, 57, 59, 61, 69, 71

3.4   1, 3, 5, 7, 9, 11, 13, 15, 17, 19, 21, 23, 25, 27, 29, 31, 33, 35, 41, 45,
       49, 51, 53, 55, 83, 85
3.5   1, 3, 17, 19, 21, 23, 25, 27, 29, 31, 33, 35, 37,39, 41,42, 43, 44, 45,
       46, 47, 48, 49, 51, 53, 55, 57, 59, 61, 63, 65
3.6   1, 3, 5, 7, 9, 11, 13, 14, 15, 16, 17, 19, 21, 23, 25, 27, 29, 1, 33, 35,
       37, 39, 41, 43, 45, 47, 49, 51, 53, 55, 57, 58,59, 60, 61, 62, 63, 64,
       65, 67, 69, 71, 73
4.1   1, 3, 5, 7, 9, 11, 13, 15, 17, 19, 21, 23, 25, 27, 29, 31, 33, 35, 37, 39,
       41, 43, 45, 47, 49, 51, 53, 55, 57, 59, 61, 63, 65, 67, 69, 71, 87, 89,
       91, 93
4.2   1, 3, 5, 7, 9, 11, 13, 15, 17, 19, 21, 23, 25, 27, 29, 31, 37, 39, 41, 43,
       45, 47, 49, 51, 53, 55, 57, 59, 61, 63, 65, 67, 69, 71, 73, 75, 77, 79,
       81, 83, 101, 102, 103
4.3   1, 3, 5, 7, 9, 11, 13, 15, 17, 19, 21, 23, 25, 27, 29, 31, 33, 35, 37, 39,
       41, 43, 45, 47, 49, 59, 61, 63, 65, 67, 69, 71, 73, 75, 77, 79, 81, 83,
       85, 87, 89, 91, 93, 95
4.4   1, 3, 5, 7, 9, 11, 13, 15, 17, 19, 21, 23, 25, 27, 31, 33, 35, 37, 39, 41,
       49, 51, 53, 55, 57, 67

Appendix B
Skills List — Fall 1999

TEXT: College Algebra and Trigonometry by Mark Dugopolski, Second
Edition. Reading: Addison-Wesley, 1999.

Component A:

- Evaluate expressions involving exponents, both integral and rational.
- Simplify expressions using rules for exponents, both integral and rational.
- Evaluate radical expressions.
- Convert expressions involving rational exponents to radical notations and vice versa.
- Simplify radical expressions.
- Do arithmetic with radical expressions.
- Do polynomial arithmetic, including FOIL, the special product rules, and long division.
- Simplify expressions by rationalizing the denominator

- Express areas and volumes of geometric object with variable dimensions.
- Factor polynomials by grouping, the special product rules, and substitution.
- Factor $ax^2 + bx + c$ by inverse FOIL method.
- Find the domain of a rational expression.
- Reduce rational expressions to lowest terms.
- Do arithmetic on rational expressions, including finding the least common denominator and simplifying complex fractions.
- Write rational expressions involving work and average speed problems.
- Solve linear equations with real coefficients, including equations involving rational expressions which are reducible to linear equations.
- Distinguish among identities, conditional equations, and inconsistent equations.
- Solve a formula for a specified variable.
- Solve word problems involving linear equations: simple interest, $d = rt$, geometrical problems, sales tax, mixing problems, work problems, and consecutive integers.
- Solve quadratic equations using the square root property, factoring, completing the square, and the quadratic formula.
- Use the discriminant to find the type and number of solutions to a quadratic equation.
- Solve word problems involving quadratic equations: supply and demand, geometric problems, $d = rt$.
- Write an inequality whose solution is a given interval.
- Solve linear inequalities; express answer in interval notation and graphically.
- Understand $\cup$ and $\cap$.
- Solve compound inequalities; express answer in interval notation and graphically.
- Solve inequalities involving absolute value; express answer in interval notation and graphically.
- Write an absolute value inequality whose solution is a given interval.
- Solve quadratic inequalities; express answer in interval notation and graphically.
- Solve rational inequalities; express answer in interval notation and graphically.
- Solve polynomial inequalities in factored form; express answer in interval notation and graphically.

Component B:

- Graph ordered pairs.
- Sketch graph of linear equations by plotting points.
- Sketch graph of linear equation by plotting intercepts.
- Use the distance and midpoint formulas.
- Use the distance and midpoint formulas to analyze geometric objects.
- Determine whether a list of points is a function.
- Determine whether an equation represents a function.
- Find the domain and range of a relation.
- Use functional (f(x)) notation.
- Find a difference quotient for a given function and simplify.
- Use geometrical relationships to define functions.
- Solve word problems involving average rate of change.
- Determine the radius and center of a circle from its equation.
- Write the standard form of the equation for a circle with a given radius and center.
- Graph a relation by plotting points.
- Determine the domain and range of a relation and whether it is a function from its graph.
- Sketch the graph of a piecewise defined function.
- Determine on what intervals a function is increasing, decreasing, or constant from its graph.
- Use transformations to sketch graphs.
- Determine the symmetry of a graph and whether a given function is even or odd.
- Solve inequalities graphically.
- Add, subtract, multiply, divide, and compose functions.
- Show that two functions are inverses.
- Determine whether a function is invertible.
- Find the inverse of an invertible function.
- Determine whether two functions are inverses of each other graphically.
- Write formulas involving direct, inverse, joint, and mixed variation.
- Find the constant of variation given the relationship between two variables and a data point.
- Solve word problems involving direct, inverse, joint, and mixed variation.
- Find the slope of a line containing two given points.
- Identify slope and y-intercept of a linear equation in various forms.

- Sketch the graph of a line using slope and intercept data.
- Find the standard form of the equation of a line through two given points.
- Find slopes and equations of lines parallel or perpendicular to given lines.
- Solve applied problems involving linear equations.
- Convert a quadratic function to the form $y = a(x - h)^2 + k$ and sketch its graph.
- Find the vertex of a quadratic function, and interpret this a maximum or minimum in applied problems.
- Find the axis of symmetry, range, intercepts, and direction of opening of a quadratic function.
- Use the graph of an appropriate quadratic function to solve a quadratic inequality.

Component C:

- Determine whether a number is a zero of a given polynomial.
- Find all real zeros of a given polynomial.
- Use the rational zero theorem to find all possible rational zeros of a given polynomial.
- Find the degree of a polynomial.
- Find the multiplicity of a zero of a polynomial.
- Solve equations involving higher-degree polynomials by factoring.
- Solve equations involving square roots of polynomials.
- Solve equations reducible by substitution to quadratic equations.
- Solve equations involving absolute value.
- Solve geometrical word problems involving any of these 3 preceding types of equations.
- Discuss the behavior of the graph of a polynomial at its x-intercepts.
- Use the leading coefficient test to determine the behavior of the graph of a polynomial as $x \to \infty$ *or* $x \to -\infty$.
- Sketch the graph of a polynomial.
- Find the domain of a rational function.
- Find the equations of the asymptotes (vertical, horizontal, oblique), for a given rational function.
- Find all intercepts of the graph of a given rational function.
- Sketch the graph of a rational function, including those not in lowest terms.

- Evaluate exponential functions.
  ketch the graph of an exponential function by plotting points.
- Use transformations to sketch graphs of functions related to exponential functions.
- Solve exponential equations by matching bases.
- Evaluate logarithmic functions.
- Sketch the graph of a logarithmic function by plotting points.
- Use transformations to sketch graphs of functions related to logarithmic functions.
- Convert exponential equations to logarithmic equations and *vice versa*.
- Solve exponential equations using logarithms.
- Solve logarithmic equations by conversion to exponential equations.
- Rewrite a sum or difference of logarithms as a single logarithm.
- Expand a single logarithm into a sum or difference of logarithms.
- Use the properties of logarithms to rewrite logarithmic expressions.
- Solve various forms of exponential and logarithmic equations.
- Solve problems involving compound interest, population growth, and radioactive decay.

References

Marcus, Nancy C. Modular Precalculus, Clustering, and Reform Mathematics. Focus on Calculus: Newsletter for the Calculus Consortium Based at Harvard University Issue 16 (1999) 1.

INDEX

CONTRIBUTORS PROFILES

Louise McNertney Berard is a Professor at Wilkes University, where she chaired the Department of Mathematics and Computer Science from 1992 to 1996. She received a B.S. in mathematics from King's College and a Ph.D. in mathematics from Brown University. She has taught a wide variety of courses in mathematics and computer science, and has made presentations locally and nationally on the use of technology to enhance student understanding of mathematical concepts. In 1988 she received the Carpenter Outstanding Teacher Award at Wilkes. She has also received outstanding performance designations on three separate occasions. Since 1980 Professor Berard has been a member of the Eastern Pennsylvania and Delaware Section of the Mathematical Association of America, serving from 1992 to 1995 as coordinator of the Section's Visiting Lecturer and Consultant Program, from 1993 to 1995 as Vice-President of the Section, and form 1995 to 1997 as Section President.

Christine L. Ebert received her Ph.D. in Curriculum and Instruction from the University of Delaware in 1994. Dr. Ebert has been a member of the Mathematical Sciences Department of the University of Delaware since 1984 and has been involved with the preparation of both elementary and secondary mathematics teachers. Her research interests include the investigation of teachers' pedagogical content knowledge, the cognitive development of the concept of function, and the analysis of the technological factors that promote conceptual understanding of functions and graphs. Dr. Ebert recently completed a year as a project director for the Delaware State Systemic Initiative, an NSF-funded collaborative for the improvement of math and science education in the schools. She currently directs teacher-networks at both the elementary and secondary level devoted to implementing investigative instructional practices in mathematics.

Philip A. DeMarois has taught mathematics and computer science at the high school, community college, and university levels for twenty-six years. He is professor of Mathematics at Mt. Hood Community College, Oregon. Dr. DeMarois is the Northwest Region vice president of AMATYC. Phil is the author of *College Algebra Laboratories Using Derive* and *TI-83 or TI-82 Mini-Labs Algebraic Investigations*. He is the co-author of *Mathematical Investigations: An Introduction to Algebraic Thinking* and *Applying Algebraic Thinking to Data*, both of which arise out of a curriculum reform project partially funded by NSF. Phil earned a Ph.D. in Mathematics Education under Dr. David Tall at the University of Warwick in Coventry, England in Spring, 1998. His research interests include the development of the function concept in students who have previously been unsuccessful learning algebra. Phil was awarded AMATYC's Teaching Excellence Award for the Midwest Region in November, 1997.

Ron C. Goolsby is professor of mathematics at Winthrop University in Rock Hill, South Carolina. He received his Ph.D. from the University of North Carolina at Chapel Hill in the area of complex variables. His interests lie both in pure mathematics and mathematics education.

Jay M. Jahangiri is an Associate Professor of Mathematics at Kent State University's Kent Campus and Geauga Campus. Jay earned his Maters Degree in mathematics from The University of London and his Doctorate in Complex Analysis from the University of York in England. Dr. Jahangiri has over fifty publications in both pedagogy of mathematics and pure mathematics that are the areas of research interest to him.

Stephen Hibberd is a senior lecturer in the School of Mathematical Sciences, University of Nottingham which he joined in 1979. His teaching experience is varied and includes a range of modules in applied mathematics and mathematical techniques to honor mathematics students and engineering students. He has been active in the areas of teaching and learning within a national, university and school framework. He is an Executive Steering Group member of the UK Mathematics Courseware Consortium that developed 'Mathwise' and co-organizer of conferences on national conferences on the 'Mathematical Education of Engineers'. Appointed as an Advisor on part-secondment part to his University's Teaching Enhancement Office since 1996 he has informed and enabling support of enhanced teaching methods throughout the University. He has also played a major role in the development and implementation of university-wide practices of providing basic IT skills to undergraduate students and the training of postgraduate students in supporting teaching and assessment activities. He is a QAA subject reviewer in Mathematics and recently received a University 'Lord Dearing Award' for his Teaching and Learning activities.

Jayne M. Kracker is the Director of Workforce Development and Continuing Studies at the Geauga Campus of Kent State University. Jayne earned her Bachelors Degree in Business Education from Kent State University and is doing graduate work at Kent State in Organization Development. Ms. Kracker is also an Adjunct Instructor in Business Technology at Kent State and has extensive experience in the business arena having worked in manufacturing and law, as well as teaching/developing Adult Education programs at area high schools. Creating university/business partnerships and researching training techniques and methods to improve the workforce are areas of interest.

Myrtle Lewin has been teaching at Agnes College since 1983. Her Ph.D. is in analysis, but in her varied experiences in both large universities (in Israel and South Africa) and at Agnes Scott, and as a parent, her interest in teaching and learning has been central. Besides, her teaching, Dr. Lewin gives workshops on collaborative learning, is an enthusiastic but novice geometer, and has a passion for forests, particularly those she has planted.

Leonard J. Lipkin, (B.A., Mathematics, Oberlin College; Ph.D., Mathematics, University of Michigan) Professor of Mathematics at the University of North Florida, is a former Chair of the Department of Mathematics and Statistics. He has published research papers in analysis and is co-author of a research monograph *Polyharmonic Functions* (Oxford University Press, 1983). He has received the Outstanding Teacher Award and the Distinguished Professor Award at the University of North Florida, and the Award for Innovative Contributions to Teaching, Learning, and Technology at the Sixth International Conference on College Teaching and Learning. As Department Chair in the 1980s he directed an effort to revise and improve the Department's course for Elementary Education majors. He was involved early in the calculus reform movement as a test-site instructor of the NSF-supported Duke Project CALC and later wrote interactive computer modules in differential equations for another NSF-supported project at Duke.

He received a three-year grant from NSF titled "Technology, Discovery, and Communication in Secondary School Mathematics." With collaborators, he has led extended projects for middle school teachers. He has been a frequent speaker at mathematics education conferences on curriculum reform, technology, and using simulation in probability and statistics. Since 1997 he has organized and directed a College Board certified AP Statistics Teacher Training Institute.

Andrew Looms is the Mathematics Learning and Technology Officer at Keele University and as such has been deeply involved with the integration of computerized diagnostic testing and mathematical courseware, *Mathwise,* into the first year undergraduate mathematics course at Keele.

Tabitha T. Y. Mingus is an assistant professor in collegiate mathematics education at Western Michigan University where she received her B.A. in Mathematics-Secondary Education. She received her Ph.D. in Educational Mathematics from the University of Northern Colorado and her M.A. in Mathematics from Central Michigan University. Mingus attributes her primary research interest of improving the mathematical training of pre-service teachers to her experiences as a middle school mathematics and science teacher. She especially enjoys working with students in post-calculus mathematics course and using technology, cooperative learning groups, and explorations to help students understand the need for abstraction and proof in mathematics.

Richard Mitchell spent three years teaching secondary mathematics education in Vail, Colorado, after graduating from Illinois State University with a Master's degree in Mathematics. In 1990, he earned a Ph.D. in mathematics education from the University of Wyoming; the dissertation dealt broadly with student misconceptions in mathematics. Currently, Dr. Mitchell is a Professor of Mathematics at the University of Wisconsin-Stevens Point where he teaches a variety of mathematics and mathematics education courses. His research interests focus on student misconceptions and the use of technology within mathematics education.

Patrick F. Mwerinde received a B.S. in Mathematics and Philosophy (double major) from Manhattan College in 1977, an M.A. in Mathematics and Computer Science from Columbia University in 1981, an M.Ed. in Mathematics and Statistics from Columbia University, Teachers College in 1983, and a Ph.D. in Mathematics from Columbia University in 1993. In 1990 I joined the faculty at the University of Delaware in the Department of Mathematical Sciences.

Thomas W. Polaski is an associate professor of mathematics at Winthrop University. He received a B.S. from Furman University and an A.M. and Ph.D. from Duke University. His areas of interest include stochastic processes, mathematical biology, and the teaching of mathematics (especially linear algebra).

Douglas Quinney has been at Keele University since 1975. He has taught all level of mathematics courses from third level option courses in Numerical Mathematics through to service course in Mathematics for Economics and Chemistry. During the last ten years he has been active primarily with the development of courseware. He is a founder member of the United Kingdom Courseware Consortium, which has developed the TLTP material called *Mathwise,* and is currently a member of the Executive Committee. He is also joint author of the award winning software packages called Calculus Connections, (European Academic Software Awards 1996, 1998), and a member of the Calculus Consortium in Higher Education, based at Arizona, which has produced a number of best selling textbooks in undergraduate Calculus. Currently, he is developing the 'Global Classroom' as part of a project based at Jacksonville University which has recently received funding from the National Science Foundation.

Thomas W. Rishel is the Associate Executive Director of the Mathematical Association of America in Washington, D.C. He moved there in 2000, leaving his former position of Senior Lecturer and Director of Undergraduate Teaching in the Department of Mathematics at Cornell University. In his current position at MAA, Dr. Rishel is involved in setting up programs and providing services for members, and for mathematicians in general.